北大社·"十四五"普通高等教育本科规划教材
高等院校机械类专业"互联网+"创新规划教材
浙江省普通高校"十三五"新形态教材

（双语教学版）

数 控 技 术

（第2版）

Computer Numerical Control Technology

（Second Edition）

主　编　吴瑞明
副主编　徐　兴　倪成员　吴蒙华
　　　　李雄兵　胡　平

内容简介

数控技术在现代制造业中占据重要地位，是实现工业自动化、柔性制造和制造信息化的基础。本书系统地介绍了数控技术概论、数控编程、CNC 装置、CNC 机床的伺服系统和位置检测装置、CNC 机床的机械结构和刀具系统、CNC 技术发展等知识，并提供参考译文，以便读者学习。

本书可作为数控技术应用、机电一体化、机械制造及其自动化、模具设计与制造等专业的数控技术双语教学用书和专业英语教学用书，也可作为数控技术等相关专业技术人员的英语参考用书。

图书在版编目（CIP）数据

数控技术：双语教学版 / 吴瑞明主编. -- 2版. 北京：北京大学出版社，2025. 1. --（高等院校机械类专业"互联网+"创新规划教材）. -- ISBN 978 - 7 - 301 - 35774 - 3

I．TP273

中国国家版本馆 CIP 数据核字第 2024QJ4868 号

书　　　名	数控技术（双语教学版）（第 2 版） SHUKONG JISHU（SHUANGYU JIAOXUE BAN）（DI - ER BAN）
著作责任者	吴瑞明　主编
策 划 编 辑	童君鑫
责 任 编 辑	关　英
标 准 书 号	ISBN 978 - 7 - 301 - 35774 - 3
出 版 发 行	北京大学出版社
地　　　址	北京市海淀区成府路 205 号　100871
网　　　址	http://www.pup.cn　新浪微博：@北京大学出版社
电 子 邮 箱	编辑部　pup6@ pup.cn　总编室　zpup@ pup.cn
电　　　话	邮购部 010 - 62752015　发行部 010 - 62750672　编辑部 010 - 62750667
印 刷 者	河北文福旺印刷有限公司
经 销 者	新华书店
	787 毫米×1092 毫米　16 开本　19.5 印张　475 千字 2017 年 3 月第 1 版 2025 年 1 月第 2 版　2025 年 1 月第 1 次印刷
定　　　价	59.80 元

未经许可，不得以任何方式复制或抄袭本书之部分或全部内容。
版权所有，侵权必究
举报电话：010 - 62752024　电子邮箱：fd@ pup.cn
图书如有印装质量问题，请与出版部联系，电话：010 - 62756370

Preface to the Second Edition

Computer Numerical Control (CNC) technology integrates computer, automatic control, information, sense, and machining technology. The report of the 20th National Congress of the Communist Party of China pointed out that we need to promote new industrialization, accelerate the construction of a strong manufacturing country, and promote the high-end, intelligent, and green development of the manufacturing industry. It is fundamental to realize automation, flexibility, and informatization in the manufacturing industry. CNC technology has become one of fundaments for advanced manufacturing technology.

This textbook incorporates feedback from front-line teachers and students, making the following modifications based on the first edition and its revision:

(1) Added reference translations to facilitate better bilingual learning for students.

(2) Chapter 6 includes introductions to intelligent manufacturing, artificial intelligence, virtual reality, big data, and other technologies.

(3) Removed the vocabulary appendix and experimental guidance.

(4) Added video and animation links related to the new form of textbooks.

This textbook is edited by Professor Ruiming Wu (Editor-in-chief, Zhejiang University of Science and Technology), with Professor Xing Xu (Zhejiang University of Science and Technology), Associate Professor Chengyuan Ni (Quzhou College), Professor Menghua Wu (Guangdong University of Science and Technology), Professor Xiongbing Li (Central South University), and Associate Professor Ping Hu (Wuhan University) serving as associate editors. Postgraduate student Xiang Lu contributed to the collection and proofreading of some Chinese materials. Thanks to Professor Joseph A. Turner (University of Nebraska-Lincoln) for providing some teaching videos.

"Numerical Control Technology (Bilingual)" has been recognized as a first-class undergraduate course in Zhejiang Province, and thistextbook has received funding from the first batch of new form textbook projects during the 13th Five-Year Plan for ordinary higher education institutions in Zhejiang Province.

Due to the limited capabilities of the editors, there may be unavoidable shortcomings in thistextbook; we welcome constructive criticism from readers.

Editors
July 2024

第 2 版前言

计算机数控（CNC）技术包括计算机技术、自动化控制技术、信息技术、传感技术和制造加工技术等。党的二十大报告指出，我们要推进新型工业化，加快建设制造强国，推动制造业高端化、智能化、绿色化发展。计算机数控技术是实现工业自动化、柔性制造和制造信息化的基础，也是先进制造技术的基础。

本书结合一线教师和学生的反馈，在第 1 版及修订版的基础上作出以下修改。

（1）增加了参考译文，以便学生更好地进行双语学习。

（2）第 6 章增加了智能制造技术、人工智能技术、虚拟现实技术、大数据技术等的介绍。

（3）删除了附录词汇和实验指导。

（4）增加了新形态教材的视频和动画链接。

本书由浙江科技大学吴瑞明教授主编，浙江科技大学徐兴教授、衢州学院倪成员副教授、广东科技学院吴蒙华教授、中南大学李雄兵教授、武汉大学胡平副教授任副主编。研究生卢翔完成部分中文资料的收集和校对。感谢内布拉斯加大学林肯分校（University of Nebraska-Lincoln）Joseph A. Turner 教授提供部分教学视频。

"数控技术（双语）"获浙江省一流本科课程，本书获浙江省普通高校"十三五"首批新形态教材项目资助。

由于编者水平有限，书中难免有不妥之处，敬请广大读者批评指正。

编者

2024 年 7 月

【资源索引】

【词汇表】

Contents

Chapter 1 Introduction of CNC 1
- 1.1 History of NC Development 1
- 1.2 Concepts of NC and CNC 2
- 1.3 Classifications of CNC Machine Tools 8
- 1.4 Characteristics of NC and CNC Machine Tools 14
- Exercises 19

Chapter 2 CNC Programming 20
- 2.1 Introduction of the CNC Programming 20
- 2.2 Basis of the CNC Programming 24
- 2.3 Definition of the CNC Programming 28
- 2.4 CNC Programming Examples 33
- 2.5 Computer Aided Manufacturing 66
- Exercises 71

Chapter 3 CNC Unit and Control Principle 76
- 3.1 Hardware Architecture of a CNC Unit 76
- 3.2 CNC System Software 83
- 3.3 Interpolation Principle 85
- 3.4 Tool Compensation Principle 87
- 3.5 CNC Acceleration and Deceleration Control 92
- 3.6 PLC Function 94
- Exercises 97

Chapter 4 CNC Machine Tool's Servo Systems and Position Measuring Devices 98
- 4.1 Introduction of CNC Machine Tool's Servo Systems 98
- 4.2 Servo Control 100
- 4.3 Position Measuring Devices 110
- Exercises 116

Chapter 5 Mechanical Construction and Tool System of CNC Machines 117
- 5.1 CNC Machine Tools and CNC Machining Centers 117
- 5.2 Mechanical System of CNC Machine Tools 124
- 5.3 Tool System of CNC Machine Tools 132
- Exercises 143

Chapter 6　CNC Technology Development　146
6.1　Open CNC System　146
6.2　STEP-NC　149
6.3　Advanced Applications of CNC Technology　150
6.4　Intelligent Manufacturing Technologies　162
Exercises　168

目 录

第1章 数控技术概论 ... 169
1.1 数控技术的发展历程 ... 169
1.2 数控技术与计算机数控技术 ... 170
1.3 数控机床的分类 ... 174
1.4 数控机床与计算机数控机床的特性 ... 179
本章小结和学习导图 ... 183

第2章 数控编程 ... 184
2.1 数控编程概述 ... 184
2.2 数控编程基础 ... 187
2.3 数控编程约定 ... 191
2.4 数控编程实训 ... 195
2.5 计算机辅助制造 ... 227
本章小结和学习导图 ... 230

第3章 CNC装置和控制原理 ... 231
3.1 CNC装置的硬件结构 ... 231
3.2 CNC系统控制软件 ... 237
3.3 插补原理 ... 238
3.4 刀具补偿原理 ... 240
3.5 CNC系统的加减速控制 ... 244
3.6 CNC系统中的PLC ... 245
本章小结和学习导图 ... 247

第4章 CNC机床的伺服系统和位置检测装置 ... 249
4.1 CNC机床伺服系统概述 ... 249
4.2 伺服控制 ... 251
4.3 位置检测装置 ... 258
本章小结和学习导图 ... 263

第5章 CNC机床的机械系统和刀具系统 ... 264
5.1 CNC车床及CNC加工中心 ... 264
5.2 CNC机床的机械系统 ... 269
5.3 CNC机床的刀具系统 ... 276
本章小结和学习导图 ... 284

第6章 CNC技术发展 ·············· 286
6.1 开放式 CNC 系统 ·············· 286
6.2 STEP-NC ·············· 288
6.3 数控技术的高级应用 ·············· 289
6.4 智能制造技术 ·············· 299
本章小结和学习导图 ·············· 303

参考文献 ·············· 304

Chapter 1
Introduction of CNC

Objectives

- To understand the history of NC development.
- To understand the concepts of NC and CNC.
- To understand the classifications of CNC machine tools.
- To understand the characteristics of NC and CNC machine tools.

1.1 History of NC Development

Numerical control is called NC for short. It is an auto control technology which has been developed in modern times and a means by which the numerical information can fulfill automatic control operation of machine tools. It minutes down in advance the machining procedure and the motion variable, such as coordinate direction steering and speed of axes on the control medium in the form of numbers, and it automatically controls the machine motion by the NC device at the same time. It also has some functions of finishing automatic tools conversion, automatic measuring, lubrication, and automatic cooling, etc.

1947 was the year in which NC was born. Because of an urgent need, Parson's Corporation, a manufacturer of helicopter rotor blades, could not make its templates fast enough, then it invented a way of coupling computer equipment with a jig borer.

In 1949, U.S. Air Force realized that parts of planes and missiles were becoming more complex. Also, the designs were constantly being improved; changes in drawings were frequently made. To speed up production, an air force study contract was given to Parson's Corporation, and the servomechanisms lab of MIT was the subcontractor.

Today, the development of NC machine tools completely depends on NC systems. NC systems have experienced two stages and six generations since USA produced the first NC milling machine in 1952.

1. NC stage (1952—1970)

The early computing speed was very low, which did not have too much effect on the scientific computing and data handling. Researchers had to set up a machine tool computer as a control system by using digital logic circuit. This stage experienced three generations.

The first generation of NC (1952—1959): Device was composed of electronic tube element.

The second generation of NC (1959—1965): Device was composed of transistor tube element.

The third generation of NC (1965—1970): Device was composed of small and medium scale integrated circuits.

【拓展视频】

2. CNC Stage (since 1970)

The small-sized general-purpose computers were mass-produced in the 1970s. Its computing speed was much higher than that in the 1950s and 1960s. These small-sized general-purpose computers were much lower in cost and much higher in reliability than the specialized computers. Therefore, they were transferred as kernel parts of NC systems and computer numerical control (CNC) stage began. With the development of computer technology, CNC stage also experienced three generations.

The fourth generation of NC (1970—1974): The small-sized general-purpose computer control system of the large scale integrated circuit was greatly applied.

The fifth generation of NC (1974—1990): The microprocessor was applied to NC systems.

The sixth generation of NC (since 1990): The NC system has entered the PC-based era.

Now, the intelligent CNC devices and intelligent machine tools are developed. They integrated intelligent sensing, internet of things, and digital twins technology.

1.2 Concepts of NC and CNC

1.2.1 NC Technology

1. Numerical Control

Numerical control (NC) is a form of programmable automation, its mechanical actions are controlled by a program containing coded alphanumerical data. The alphanumerical data represents relative positions between a workhead and a workpiece, as well as other instructions needed to operate the machine tool. The workhead is a cutting tool or other processing apparatus, and the workpiece is the object being processed. When the current job is completed, the program of instructions can be changed to process a new job. The capability to change the program makes NC suitable for low and medium productions.

It is much easier to write new programs than to make major alterations of the processing equipment.

2. Basic Components of NC Machine Tools

The control systems of NC machine tools can handle many tasks that commonly completed by the operators of conventional machine tools. NC systems must know when and in what sequence it should issue commands to change tools, at what speeds and feeds the machine tools should operate, and how to work a part to the required size. The system gains the ability to perform control functions through the numerical input information: the control program.

A NC machine tool has five basic components: input media, machine control unit, servo-drive unit, feedback transducer, and mechanical machine tool unit (Figure 1.1).

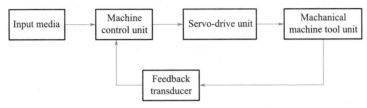

Figure 1.1 Basic Components of NC Machine Tool

The work process of NC machine tool is shown in Figure 1.2. First of all, NC programmers should study the part drawing and the process chart, and then prepare the control program in a specified format as it contains all the necessary control information. A computer-assisted NC programming for NC machine tool is also available, in which the computer considerably facilitates the work of programmers and generates a set of NC instructions. Next, the NC program is transferred to a control computer. A widely-accepted method is that programmers input the NC program via a keyboard on the operation panel. The computer converts each command into a signal that the servo-drive system needs. Then, the servo-drive system then drives the machine tool to manufacture the finished part.

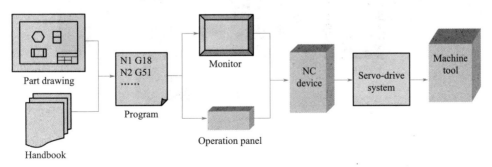

Figure 1.2 The Work Process of NC Machine Tool

1.2.2 CNC Technology

1. CNC system

In the 1970s, CNC machine tools were developed with minicomputers being used as control units. With the advances in electronics and computer technologies, current CNC systems employ several high-performance microprocessors and programmable logical controllers (PLCs) that work in a parallel and coordinated mode.

A CNC machine tool addes an onboard computer on the basis of a NC machine tool. The onboard computer is often referred to as the machine control unit (MCU). MCUs for NC machine tools are usually "hard" wired, which means all the machine functions are controlled by physical electronic elements that are built into the controllers. The onboard computer is "soft" wired, which means the machine functions are encoded into the computer at the time of manufacturing, and they will not be erased when the NC machine tools are turned off. The computer memory that holds such information is known as the read-only memory (ROM).

The MCU usually has an alphanumeric keyboard for direct data input or manual data input (MDI) of programs. Such programs are stored in the random access memory (RAM) and can be played back, edited, and processed. All the programs residing in the RAM, however, are lost when the CNC machine tools are turned off. These programs can be saved on auxiliary storage devices, such as punched tapes, magnetic tapes, and magnetic disks. New MCUs have graphic monitors that can display not only CNC programs but tool paths and any errors in the programs.

2. Components of CNC System

A CNC machine tool is composed of the following parts (Figure 1.3):

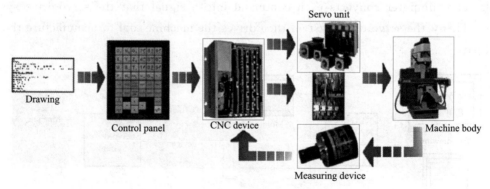

Figure 1.3 Components of CNC Machine Tool

(1) CNC device.

The CNC device is the kernel of the CNC system. It can handle input machining programs, operation commands, and output control commands, then finish the work which the machining programs and operations need. CNC device mainly consists of computer system, position control panel, PLC interface panel, communication interface panel, extension function template, and appropriate control software. The display unit serves as an interactive device between the machine tool and the operator. When the machine tool is running, the display unit displays the present state, such as the position of the machine slide, the spindle speed, the feed rate, the machining program, etc.

In an advanced CNC machine tool, the display unit can show the graphics simulation of the tool path so that the program can be verified before the actual machining. Other important information about the CNC system (such as machine parameters, logic diagrams of the program controller, error messages, and diagnostic data) can also be displayed during the maintenance and installation operations.

(2) Servo-drive device and measure device.

Servo-drive device includes spindle servo-drive system, spindle motor, feed servo-drive system, and feed motor. Measure device refers to the position and speed measure device. It is a necessary device to finish the spindle control and closed-loop control for the feed speed and for the feed position. The spindle servo-drive system can complete the cutting motion and control the speed. The feed servo-drive system can finish the shaping motion by utilizing the parameters of feed speed and feed position. The characteristic is to sensitively and accurately find the position of the CNC system and the speed command.

(3) Control panel.

Control panel, also called operation panel, is a tool used for exchanging information between the operator and the CNC machine tool. The operator can operate, program and debug the CNC machine tools or set and alter machining parameters. The operator can also understand and inquire the motion condition of the CNC machine tool by using the control panel. It is an input/output device.

(4) Control medium and input/output equipment.

The control medium is an agent to record machining programs, and it is also a medium to set up contraction between operators and machine tools. Input/output equipment is the device by which the information exchange can be done between the CNC system and external equipment. It inputs the machining programs recorded on the control medium into the CNC system, and stores and records the debugged machining programs on the appropriate medium with the output device. Today, the control medium of CNC machine tools and the input/output equipment are disks and disk drivers.

(5) Machine body.

The machine body of CNC system is an executive part to fulfill the machining parts. It is composed of main motion parts, feed motion parts, bearing rack, automatic platform change system, automatic tool changer (ATC), and accessory devices.

3. Input Media

The input media contain instruction programs, which include detailed commands that direct actions of the machine tool. The commands refer to positions of tools relative to the worktable on which the workpiece is fixed. Additional instructions are usually included, such as spindle speed, feed rate, tool selection, and other functions. The program is coded on a suitable medium for submission to the MCU. For many years, the common medium has been punched tape, using a standard format that could be interpreted by the MCU. Nowadays, punched tape has largely been replaced by newer storage technologies, such as magnetic tape, floppy disk, and computer network transmission.

(1) Floppy disk.

A floppy disk is a small magnetic storage device for CNC data input. It is one of the most common storage media since the 1970s in terms of data transfer speed, reliability, storage size, data handling, and the ability to read and write. Furthermore, the data within a floppy disk could be easily edited at any point as long as the operator has the proper program to read it. However, it has been proven to be quite problematic in the long run as it has a tendency to degrade alarmingly fast and is sensitive to large magnetic fields, as well as the dust and scratches that usually existed in the work shop.

Figure 1.4 A USB Disk

【拓展图文】

(2) USB disk.

A USB disk (Figure 1.4) is a removable, rewritable, and portable hard drive with smaller size and larger storage than a floppy disk. It is commonly used for machine tools that do not have enough memory or storage buffer for large CNC programs.

(3) A serial communication port.

The data transfer between a computer and a CNC machine tool is often accomplished through a serial communication port. International standards for serial communication ports are established so that information can be exchanged in an orderly way. One of the most common interfaces between computers and CNC machine tools is referred to the EIA (Electronic Industries Association) standard RS-232 port. Most of PCs and CNC machine tools have built in RS-232 port, and a standard RS-232 cable is used to connect a CNC machine tool to a PC, which enables the data transfer in a reliable way. Programs can be downloaded into the memory of a machine tool or uploaded to the computer for temporary storage by running a communication program on the computer and setting up the machine controller to interact with the communication software.

Direct Numerical Control (DNC, Figure 1.5) is referred to a system connecting a set of CNC machine tools to a common memory for CNC program storage or machining program storage with provision for on-demand distribution of data to the machine tools. The CNC programs are downloaded a section at a time into the machine controller. Once the downloaded section is executed, the section will be discarded to leave room for other sections. This method is commonly used for machine tools that do not have enough memory or storage buffer for large CNC programs.

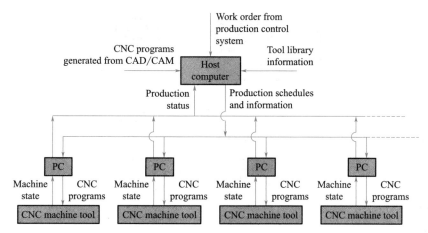

Figure 1.5   Direct Numerical Control

DNC is a hierarchical system for distributing data between a production management computer and CNC systems. The host computer is linked with a number of CNC machine tools, or computers are directly connected with the CNC machine tools for downloading programs. The communication program in the host computer can utilize two-way data transfer features for production data communication, including production schedule, parts produced, and machine utilization, etc.

(4) Ethernet communication.

Due to the advancement of computer technologies and the drastic reduction of costs of computers, it is becoming more practical and economical to transfer programs between computers and CNC machine tools via an Ethernet communication cable. This medium provides a more efficient and reliable means in programs transmission and storage. Most companies now have built a LAN (local area network) as their infrastructure, and more CNC machine tools provide an option of the Ethernet card for direct communication within the LAN.

(5) Conversational programming.

Programs can be input to the machine controller via a keyboard. Built-in intelligent software inside the machine controller enables the operator to enter the required data step by step. This is a very efficient way for preparing programs for relatively simple workpieces involving up to 2.5-axis machining.

4. CNC Manufacturing Process

The main stages involved in producing a component on a CNC system are shown in Figure 1.6.

(1) A program is written through using G-code and M-code. It describes the sequence of operations that the machine tool must perform in order to manufacture the component.

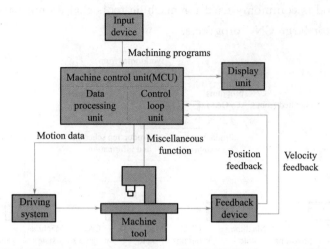

Figure 1.6 CNC Manufacturing Process

The program can be produced off-line, i.e., away from the machine tool, either manually or with the aid of a CAD/CAM system.

(2) The program is loaded into the MCU. At this stage, the program can still be edited or simulated using a keyboard or an input device.

(3) The MCU processes the program and sends signals to the components, directing the machine tool to complete the required sequence of operations.

The application of CNC to a manual machine tool allows its operation to become fully automated. Combining the CNC with the use of a program, the CNC machine tool performs repeat tasks with high degrees of accuracy.

1.3　Classifications of CNC Machine Tools

CNC machine tools are classified in different ways as follows:

(1) Types of CNC machine tools application.

(2) Types of CNC motion control system.

(3) Types of servo-drive system.

1.3.1　Types of CNC Machine Tools Application

CNC machine tools are widely used in the metal-cutting industry. They are suitable to produce the following types of parts:

(1) Parts with complicated contours.

(2) Parts requiring close tolerance and good repeatability.

(3) Parts requiring expensive jigs and fixtures if produced on conventional machine tools.

(4) Parts that have several engineering changes, i.e., parts that are during the

development stage of a prototype.

(5) In cases where human errors could be extremely costly.

(6) Parts that are needed in a hurry.

(7) Small batch lots or short production runs.

Some common types of CNC machine tools and instruments used in industry are as follows:

(1) Numerical control bender (Figure 1.7).

(2) Drilling machine.

(3) Lathe/Turning center.

(4) Milling/Machining center.

(5) Turret press and punching machine.

(6) Electro-discharge machine (EDM) (Figure 1.8).

(7) Grinding machine.

(8) Flame-cutting machine.

(9) Laser-cutting machine (Figure 1.9).

(10) Ultra-high-pressure waterjet cutting unit (Figure 1.10).

(11) Electrochemical machine tool.

(12) Laser measuring machine (Figure 1.11).

(13) Coordinate measuring machine.

(14) Industrial robot.

【拓展视频】

【拓展视频】

Figure 1.7　Numerical Control Bender

Figure 1.8　Electro-discharge Machine

Figure 1.9　Laser-cutting Machine

Figure 1.10　Ultra-high-pressure Waterjet Cutting Unit

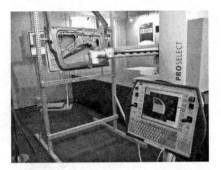

Figure 1.11 Laser Measuring Machine

1.3.2 Types of CNC Motion Control System

Some CNC machine tools (e.g., drilling, punching and spot welding) are performed at discrete locations on the workpiece. Others (e.g., turning, milling and con-tinuous arc welding) are carried out while the workhead is moving. When the tool is moving, it may be required to follow a straight-line path, a circular, or other curvilinear paths. These different types of movement are accomplished by the CNC motion control system.

CNC motion control systems can be divided into point-to-point control system, straight-cut control system, and contouring control system.

1. Point-to-point Control System

The point-to-point control system (Figure 1.12), also called the positioning control system, moves the worktable to a programmed location without regard for the path taken to get to the location. Once the movement has been completed, some processing actions are accomplished by the workhead at the location, such as drilling or punching a hole. Thus, the program consists of a series of point locations at which operations are performed.

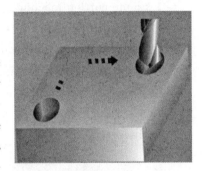

Figure 1.12 Point-to-point Control System

The MCU in a point-to-point control system contains registers that hold the axis motion commands. In some systems, the X-axis command is satisfied initially, followed by Y-axis and Z-axis commands. This operation may produce a zigzag path that will ultimately terminate at the proper point location.

【拓展视频】

2. Straight-cut Control System

The straight-cut control systems (Figure 1.13) contains a more complex MCU. In these servos, positioning commands are evaluated simultaneously so that vector motion in two axes is possible. However, the vector motion is limited to a one-to-one pulse output. Therefore, only 45° vectors maybe traced.

3. Contouring Control System

The contouring facility enables a CNC machine tool to follow any path at any prescribed feed rate. The contouring control system (Figure 1.14), also called the continuous path control system, manages the simultaneous motion of the tool in two, three, four, or five axes (the fourth and the fifth axes are angular orientations) by interpolating the proper path between prescribed points.

Figure 1.13 Straight-cut Control System

Figure 1.14 Contouring Control System

When the contouring control system is utilized to move the tool parallel to only one of the major axes of the machine tool worktable, this is called straight-cut CNC. When the contouring control system is used for simultaneous control of two or more axes in machine operations, the term contouring is used. All the contouring control systems have the ability to perform linear interpolation and circular interpolation.

4. Interpolation

One of the important aspects of the contouring control system is interpolation. The paths that a contouring control system is required to generate often consist of circular arcs and other smooth nonlinear shapes. Some of these shapes can be defined mathematically by relatively simple geometric formulas, whereas others cannot be mathematically defined except by approximation. In any case, a fundamental problem in generating these shapes using CNC equipment is that shapes are continuous, and CNC is digital. To cut along a circular path, the circle must be divided into a series of straight-line segments that approximate the circular path. The CNC machine tool is commanded to machine each straight-line segment in succession, so that the machined surface closely matches the desired shape. The maxium error between the nominal (desired) surface and the actual (machined) surface can be controlled by the lengths of the individual line segment.

If the programmer was required to specify the endpoints for each straight-line segment, the programming task would be extremely arduous and fraught with errors. Also, the program would be extremely long because of a large number of points. To ease the burden, interpolation routines have been developed to calculate the intermediate points to be followed by the tool to generate a particular mathematically defined or approximated path. A number of interpolation methods are available to deal with the various problems encountered in generating a smooth continuous path, including:

(1) The linear interpolation.

(2) The circular interpolation.

(3) The helical interpolation.

(4) The parabolic interpolation.

(5) Cubic spline interpolation.

Each of these procedures permits the programmer to generate machine instructions for linear or curvilinear paths using relatively few input parameters. The interpolation module in the MCU performs the calculation and directs the tool along the path. In the CNC system, the interpolation is generally accomplished by software. The linear interpolation and the circular interpolation are almost always included in modern CNC systems, whereas the helical interpolation is a common option. The parabolic interpolation and the cubic spline interpolation are less common, and they are only needed when producing complex surface contours.

1.3.3 Types of CNC Servo-drive System

As the actual velocity and position detected from a sensor are fed back to a control circuit, the servo motor used in the CNC machine tool is continuously controlled to minimize the velocity error or the position error. The feedback control system consists of three independent control loops (Figure 1.15) for each axis of the machine tool: the outermost control loop (a position-control loop), the middle loop (a velocity-control loop), and the innermost loop (a current-control loop). In general, the position-control loop is located in the CNC, and the others are located in a servo-drive device. However, there is no absolute standard about the location of control loops, and the locations can be varied based on the intention of the designer.

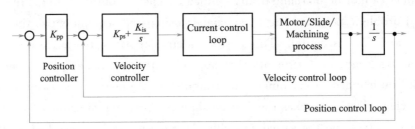

Figure 1.15 The Control Loops in CNC

In the spindle system of CNC machine tools, feedback control of velocity is applied to maintain a regular rotation speed. The feedback signal is generally generated in two ways: a tacho-generator, which generates induction voltages (analog signals) as a feedback signal, and an optical encoder, which generates pulses (digital signals). In recent times, it is typical that the feedback control is performed based on optical encoder signals instead of tacho-generator signals.

1. Open-loop Servo-drive System

Open-loop servo-drive system has no access to the real time data about the performance of the system, so there is no immediate corrective action can be taken in case of the disturbance of system. This system is normally applied only to the case where the output is almost constant and predictable. Therefore, it is unlikely to be used to control CNC machine tools since the cutting force and loading of a CNC machine tool is never constant. The only exception is the wire cut machine, for which there is virtually no cutting force in the wire cut machine.

2. Semi-closed-loop Servo-drive System

A semi-closed-loop servo-drive system uses feedback measurements to ensure that the worktable is moved to the desired position. It is characterized as a system that the indirect feedback monitors the output of servo motor. Although this method is popular within CNC systems, it is not as accurate as direct feedback. The semi-closed-loop servo-drive system compares the command position signals with the drive signals of the servo motor.

In operation, the semi-closed-loop servo-drive system is directed to move the worktable to a specified location as defined by a coordinate value in the Cartesian system. Most positioning systems have at least two axes with a control system for each axis, but the diagram only illustrates one of these axes. A servomotor connected to a lead screw is a common actuator for each axis. A signal indicating the coordinate value is sent from the MCU to the servo motor that drives the lead screw, whose rotation is converted into linear motion of positioning table. As the table moves closer to the desired coordinate value, the difference between the actual position and the input value is reduced. The actual position is measured by a feedback sensor, which is attached to servo motor axis or lead screw. This system is unable to sense backlash or lead screw windup due to varying loads, but it is convenient to adjust and has a good stability.

The semi-closed-loop servo-drive system is one of the most popular control mechanisms. In this system, a position detector is attached to the shaft of a servo motor and detects the rotation angle. The position accuracy of the axis has a great influence on the accuracy of the ball screw shafts. For this reason, ball screw shafts with high accuracy are developed and widely used. Due to the precision of ball screw shafts, the problem with accuracy has been overcome practically.

If necessary, the pitch-error compensation and backlash compensation can be used in CNC systems to increase the positional accuracy. The pitch-error compensation is that, at the specific pitch, the instructions to the servo-drive system are modified in order to remove the accumulation of positional error. The backlash compensation is that, whenever the moving direction is changed, additional pulses corresponding to the amount of backlash are sent to the servo-drive system. Recently, the usage of the ball screws with large lead high-pitch in high-speed machining has been increasing.

3. Closed-loop Servo-drive System

In the closed-loop servo-drive system (Figure 1.16), feedback devices closely monitor the output, and any disturbance will be corrected in the first instance. Therefore, high accuracy is achievable. This system is more powerful than the open-loop servo-drive system, and it can be applied to the case where the output is subjected to frequent changes. Nowadays, almost all CNC machine tools use this system.

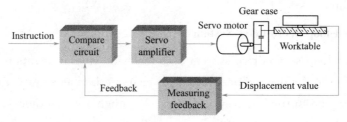

Figure 1.16　Closed-loop Servo-drive System

A semi-closed-loop servo-drive system or a closed-loop servo-drive system uses conventional variable-speed AC or DC servo motors (servo-drive system) to drive the axes.

1.4　Characteristics of NC and CNC Machine Tools

1.4.1　Advantages and Disadvantages

1. NC Machine Tools

The advantages generally attributed to NC, with emphasis on the applications of NC machine tools, are the following:

(1) Greater accuracy and repeatability. Compared with manual production methods, NC machine tools reduce or eliminate variations that are due to operators' skill differences, fatigue, and other factors attributed to inherent human variability. Parts are made closer to nominal dimensions, and there is less dimensional variation among parts in the batch.

(2) More complex part geometries are possible. NC technology has extended the range of possible part geometries beyond what is practical with manual machining methods. This is an advantage in the product design in several ways. More functional features can be designed into a single part to reduce the total number of parts in the product and in the associated assembly cost. Mathematically defined surfaces can be fabricated with high precision. The space is expanded within which the designer's imagination can wander to create new parts and product geometries.

(3) Non-productive time is reduced. The NC machine tool cannot optimize the metal cutting process, but it does increase the proportion of time. When the NC machine tool is

cutting metal, it can reduce the workpiece handling time and carry out automatic tool changes. This advantage saves labor cost and lowers elapsed time to produce parts.

(4) Lower scrap rates. Because greater accuracy and repeatability are achieved, and human errors are reduced during production, more parts are produced within tolerance. As a consequence, a lower scrap allowance can be planned into the production schedule, so fewer parts are made in each batch with the result that production time is saved.

(5) Inspection requirements are reduced. Less inspection are needed when NC machine tools are used because parts produced from the same NC programs are virtually identical. Once the programs have been verified, there is no need for the high level of sampling inspection when parts are produced by conventional manual methods. Except for the tools wear and equipment malfunction, NC machine tools produce exact replicates of the part in each machining cycle.

(6) Engineering changes can be accommodated more gracefully. Instead of making alterations in a complex fixture, revisions are made in the NC programs to accommodate the engineering changes.

(7) Simpler fixtures are needed. NC machine tools require simpler fixtures because accurate positioning of tools is accomplished by the NC machine tools, and the positioning of tools does not have to be designed in the fixtures.

(8) Shorter manufacturing time. Jobs can be set up more quickly and fewer setups are required per part when NC machine tools are used. This results in shorter elapsed time between order release and completion.

(9) Reduced parts inventory. Because fewer setups are required, and jig changeovers are easier and faster, NC machine tools permit production of parts in smaller lot sizes. The economic lot size is lower in NC machining than in conventional batch production, so the average parts inventory is reduced.

(10) Less floor space required. This results from the fact that fewer NC machine tools are required to perform the same amount of work compared with the number of conventional machine tools needed. Reduced parts inventory also contributes to lower floor space requirements.

(11) Operators' skill requirements are reduced. The skill requirements for operating NC machine tools are generally less than those required to operate conventional machine tools. The NC machine tool usually consists of loading parts, unloading parts, and periodically changing tools. The machining cycle is carried out under programs control. Tools changing for some NC machine tools can even be carried out by programs control. Performing a comparable machining cycle in a conventional machine tool requires much more participation by the operator, and a higher level of training and skills are needed.

However, the disadvantages of NC machine tools include the following:

(1) Higher investment cost. A NC machine tool has a higher first cost than a conventional machine tool for the following reasons:

① NC machine tools include NC devices and electronic hardware.

② Software development costs of the NC devices must be included in the cost of the equipment.

③ More reliable mechanical components are generally used in NC machine tools.

④ NC machine tools often possess additional features (such as ATCs and part changers) that are not included in conventional machine tools.

(2) Higher maintenance efforts. In general, NC equipment requires a higher level of maintenance than conventional equipment, which means higher maintenance and repair costs. This is due largely to the computer and other electronics that are included in modern NC systems. The maintenance staff must include the persons who are trained in maintaining and repairing this type of equipment.

(3) Parts programming. NC equipment must be programmed. To be fair, it should be mentioned that process planning must be accomplished for any parts, whether it is produced on NC equipment or not. However, NC programming is a special preparation step in batch production that is absent in conventional machine work shop operations.

(4) Higher utilization of NC equipment. To maximize the economic benefits, aNC machine tool must be operated multiple shifts. This might mean adding one or two extra shifts to the normal operations, with the requirement for supervision and other staff support.

2. CNC Machine Tools

CNC machine tools open up new possibilities and advantages that are not offered by NC machine tools.

(1) Reductions in the hardware are necessary in order to add a function of a machine tool. New functions can be programmed into MCUs as software.

(2) The CNC programs can be written, stored, and executed directly in CNC machine tools.

(3) Any portion of an entered CNC program can be played back and edited at will. Tool motions can be electronically displayed upon playback.

(4) Different CNC programs can be stored in the MCUs.

(5) Several CNC machine tools can be linked to a host computer. Programs written via the host computer can be downloaded to any CNC machine tools in the network. This is known as DNC (distribute numerical control).

(6) Several DNC systems can also be networked to form a large distributive CNC system.

(7) The CNC program can be input from USB disks and floppy disks, or downloaded from LAN.

CNC machine tools can dramatically boost productivity. The CNC manager, however, can only ensure such gains by first addressing several critical issues as the following:

(1) Sufficient capital must be allocated for purchasing advanced CNC equipment.

(2) CNC equipment must be maintained on a regular basis by obtaining a full-service contract or by hiring in-house technicians.

(3) Personnel must be thoroughly trained in the operation of CNC machine tools. In particular, many jobs require setups for machining parts to comply with tolerances of form and position.

(4) Careful production planning must be studied because the hourly cost of operation of a CNC machine tool is usually higher than that of a conventional machine tool.

1.4.2 Financial Rewards of CNC Machine Tools Investment

Investors are encouraged to look to CNC machine tools as a production solution with the following benefits:

(1) Savings in the direct labor. One CNC machine tool's output is commonly equivalent to that of several conventional machine tools.

(2) Savings in the operator training expenses.

(3) Savings in the work shop supervisory costs.

(4) Savings in tighter and more predictable production schedules.

(5) Savings in the real estate, as fewer CNC machine tools are needed.

(6) Savings in the power consumption, as CNC machine tools produce parts with a minimum of motor idle time.

(7) Savings in the improved cost estimation and pricing.

(8) Savings due to the elimination of construction of precision jigs, reduced need for special fixtures, and reduced maintenance and storage costs of these items.

(9) Savings in the tool engineering/design and documentation. The CNC machining capability eliminates the need for special tool, special boring bars, special thread tool, etc.

(10) Reduced inspection time due to CNC machine tools' ability to produce parts with superior accuracy and repeatability. In many cases, only spot-checking of critical areas is necessary without loss of machining time.

The payback period is used to estimate the investment efficiency. The payback period calculation estimates the number of years required to recover the net cost of CNC machine tools.

$$\text{Payback Period} = \frac{\text{Net Cost of CNC} - \text{Net Cost of CNC} \times \text{Tax Credit}}{\text{Savings} - \text{Savings} \times \text{Tax Rate} + \text{Yearly Depreciation of CNC} \times \text{Tax Rate}}$$

(1.1)

The return on investment (ROI) is used to estimate investment efficiency. The ROI calculation predicts the percent age of the net cost of CNC machine tools which will be recovered each year. The ROI calculation accounts for the useful life of the CNC machine tools.

$$\text{ROI} = \frac{\text{Average Yearly Savings} - \text{Net Cost of CNC/Year of Life}}{\text{Net Cost of CNC}} \tag{1.2}$$

Given the investment example (Table 1-1) for implementing a CNC machine tool, determine the payback period and the ROI. The CNC machine tool is conservatively estimated to have a useful life of 12 years.

Table 1-1 Financial Rewards of a CNC Machine Tool Investment

Initial investment ($)	130,250
One-time savings in tooling ($)	35,000
Net cost of the CNC machine tool ($)	95,250
Average yearly savings ($)	63,100
Tax credit (10%)	0.1
Tax rate (46%)	0.46
Yearly depreciation of the CNC machine tool ($)	10,900

$$\text{Payback Period} = \frac{95,250 - 95,250 \times 0.1}{63,100 - 63,100 \times 0.46 + 10,900 \times 0.46} \approx 2.19 (\text{years}) \tag{1.3}$$

This calculation estimates that the net cost of the CNC machine tool will be recovered in 2.19 years.

$$\text{ROI} = \frac{63,100 - 95,250 \div 12}{95,250} \times 100\% \approx 57.9\% \tag{1.4}$$

This calculation estimates that the investor can expect 57.9% of the net cost of the CNC machine tool (or 0.579 × $95,250 = $55,149.75) to be recovered each year if the CNC machine tool's useful life is 12 years.

1.4.3 Reliability of CNC Machine Tools

Reliability is defined as the probability that a device will perform its required function under stated conditions for a specific period of time. The life of a population of units can be divided into three distinct periods. Figure 1.17 shows the reliability "bathtub curve", which models the relationship between the instantaneous failure rates and the time. The first period is also called the infant mortality period. The next period is called the useful life (normal life). The third period (end of life) begins at the point where the slope begins to increase and extends to the end of the graph. This is what happens when the units become old and begin to fail at an increasing rate.

Mean time between failure (MTBF) is a reliability term used to provide the amount of failures per million hours for a product. This is the most common inquiry about a product's lifespan, and it is important in the decision-making process of the end user. MTBF is the predicted elapsed time between inherent failures of a system during operation. MTBF can be calculated as the arithmetic mean (average) time between failures of a system.

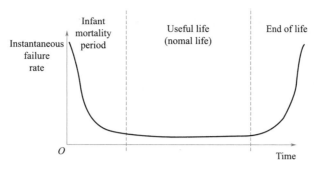

Figure1.17 The Reliability "Bathtub Curve"

$$\text{MTBF} = T/R \tag{1.5}$$

Where, T—total time,

R—the number of failures.

Mean time to repair (MTTR) is the time needed to repair a failed hardware mod-ule. In an operational system, repair generally means replacing a failed hardware part. Thus, MTTR could be viewed as the mean time to replace a failed hardware module. Repairing a product drives up the cost of the installation in a long run. To avoid MTTR, many companies purchase spare products so that the replacement can be installed quickly.

MTTR is a basic measurement of the maintainability of repairable items. It represents the average time required to repair a failed component or device. Expressed mathematically, it is the total corrective maintenance time divided by the total number of corrective maintenance actions during a given period of time.

$$\text{MTTR} = T/N \tag{1.6}$$

Where, T—total corrective maintenance time,

N—the number of units under test.

Exercises

1. Narrate the concept of NC.

2. Narrate the concept of CNC.

3. What are the components of a CNC machine tool? What is the function of each component?

4. What is point-to-point control system? What is contouring control system?

5. What is open-loop servo-drive control, semi-closed-loop servo-drive control, and closed-loop servo-drive control?

6. What are the advantages and disadvantages of CNC machine tools?

7. Narrate the financial rewards of CNC machine tools investment using ROI.

Chapter 2
CNC Programming

Objectives

- To understand the introduction of the CNC programming.
- To understand the basis of the CNC programming.
- To understand the definition of the CNC programming.
- To master the CNC programming examples.
- To understand the computer aided manufacturing.

2.1 Introduction of the CNC Programming

The CNC programming invdves creating a list of coded instructions which describes how the designed component or part will be manufactured. These coded instructions are called data, composed of a series of letters and numbers. The CNC programming includes all the geometrical and technological data to perform the required functions and movements control of the machine tool to manufacture the part.

The CNC programming is structured in a way that allows it to be broken down into separating lines of data, and each line describes a particular set of machining operations. These lines, which run in sequence, are called blocks. A block of data contains words which are sometimes called codes. Each word refers to a specific cutting/movement command or a machine function.

The programming language recognized by the CNC machine tool controller is an ISO code, which includes G-codes and M-codes. Each program word is composed of a letter called the address, along with a number.

The CNC program can contain a number of separate programs, which together describe all the operations required to manufacture the part. The main program is the controlling program, i.e, the program first read, or accessed, when the entire machining program sequence is run. This controlling program can call a number of smaller programs into operations.

These smaller programs, called subprograms, are generally used to perform repeating tasks before returning back to the main program. Normally, the controller operates according to one program. In this case, the main program is also the machining program. The main program is written by using ISO address codes.

The program of instructions is the detailed step-by-step commands that direct the actions of CNC the processing equipment. In the applications of CNC machine tools, the program of instructions is called a machining program, and the person who prepares the program is called a CNC programmer. CNC is a form of automatically operating machine tool based on coded alphanumeric data. A complete set of coded instructions for executing an operation is called a program. The program is translated into corresponding electrical signals for input to servo motors that run the CNC machine tool.

The CNC programming controls the whole process, from part graphics to the CNC manufacturing. When using CNC machine tools to manufacture parts, the CNC programming is very important. The CNC programming is required to be not only correct and fast but to be effective and economical.

2.1.1 The Contents and Steps of CNC Programming

【拓展视频】

Before CNC programming, the CNC programmer should understand the specifications and characteristics of the CNC machine tool, the functions and programming instruction format of the CNC system, etc. When programming, the CNC programmer should analyze the part's technical requirements, geometrical shape dimensions, and technological requirements. Then the CNC programmer can determine the manufacturing methods, calculate numerical values, and get tool positions. According to part dimensions, tool position values, cutting parameters (i.e., spindle speed, feed rate, cutting depth), and auxiliary functions (ATC, CW, CCW, coolant on/off), the CNC programmer can program. The program can be input into the CNC system, and the CNC system controls CNC machine tools to manufacture automatically.

Typically, the CNC programmer follows a certain processor workflow that can be summarized into several critical steps: analyzing part graphics, determining the manufacturing technological process, selecting the program origin and the coordinate system, calculating numerical values, writing programs, verifying programs, and inputting programs into the CNC system.

(1) Analyzing part graphics and determining the manufacturing technological process.

This step includes analyzing part graphics, understanding the machining contents and requirements, determining technological processes, machining plans, machining sequence, machining routes, fixing methods, cutting parameters and selecting suitable cutting tools, etc. Besides, the CNC machine tools' codes should be understood clearly and the CNC machine tools' functions should be exploited fully.

(2) Selecting the program origin and the coordinate system.

In the manufacturing process of CNC machine tools, it is important to correctly select the program origin and the coordinate system. In the CNC programming, the program coordinate system is the standard coordinate system ascertained on the workpiece.

(3) Calculating numerical values.

This step is to get the tool path according to the part geometric dimension and the method of cutter radius compensation, thus obtaining the cutter position.

(4) Writing programs.

After determining the machining route, the technological process and the coordinate values of the tool path, step-by-step, the CNC programmer can write programs in accordance with the specified function codes and program formats of the CNC system.

(5) Verifying programs.

Before the program is used in the real production, the program must be checked. The tool path errors that could ruin the part, damage the fixtures, break the cutting tool or crash the machine must be detected. In some cases, the program can be tested through manufacturing a part on a CNC machine tool. On the basis of detecting results, the program is needed to be modified and to be adjusted until the program satisfies the machining requirements completely.

(6) Inputting programs into the CNC system.

It is an unpleasant reality that many CNC programs, regardless of how they were developed, lack the background information that can help the machine operator. At best, the program may include some basic data regarding the setup and even some special instructions. The operator needs to know what fixture has been used, how the part is oriented, what tools have been selected, and where the part is located.

The program is coded on a suitable medium for submission to the MCU. For many years, the common medium was the punched tape using a standard format that could be interpreted by the MCU. Today, the punched tape has been replaced by newer storage technologies in modern work shops. These technologies include magnetic tapes, diskettes, USB flashes, and electronic transfer of programs from a computer.

The steps above-mentioned are programmed manually. This programming method is called manual programming. A CNC programmer must not only have the knowledge of structures of machine tools, the functions and standards of the CNC system, but also have the knowledge of technological processes, such as fixtures, tools, cutting parameters, etc.

Keep in mind that this is not always a step-by-step method as it may appear to be. Often, a decision made in one step influences a decision made in another step, which often leads to revisiting earlier stages of the process and making necessary changes.

Once the part graphics and programs reach the machine work shop, it is up to the CNC operator to continue with the actual production, but the production cannot start right away. The following workflow needs to be completed:

(1) Evaluate the program.

(2) Check supplied materials.

(3) Prepare required tools.

(4) Set up and register tools.

(5) Set up the part in a fixture.

(6) Load the program.

(7) Set various offsets.

(8) Run the first part.

(9) Optimize the program if necessary.

(10) Run the production.

(11) Check the program frequently.

【拓展视频】

【拓展视频】

2.1.2 The Methods of the CNC Programming

The CNC programming can be accomplished by using a variety of methods ranging from manual methods to highly automated methods.

1. The Manual Programming

The CNC programming in which the whole programming is completed manually is called the manual programming, including calculating numerical values on a computer.

【拓展视频】

In many mechanical manufacturing trades, there are a large number of uncomplicated parts that are constituted only by the simple geometric elements of straight lines and circles. The numerical values of the parts are calculated simply. The blocks of a program are not numerous, and checking the program is easy. These programs can be completed manually, so the manual programming is still a very common programming method at home and abroad.

However, the manual programming has difficulty in programming complicated parts that have non-circular curves and surfaces. Therefore, the automatic programming is required when programming.

The procedures of the manual programming (Figure 2.1) can be divided into four steps:

(1) Analyze workshop drawings.

(2) Define work plans.

(3) Select clamping devices and necessary tools (setup sheets).

(4) Write CNC programs (program sheets).

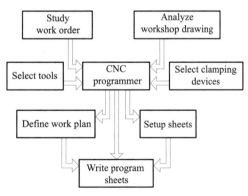

Figure 2.1 Procedures of the Manual Programming

Various documents must be analyzed, and plans for production execution must be created.

2. The Automatic Programming

The automatic programming is also called computer-aided programming. Most or all of the programming is completed by a computer, such as calculating numerical values, writing programs, fabricating control media, etc. The automatic programming lightens the programming intensity of CNC programmers, shortens the programming time, and improves the programming quality. At the same time, it solves the complicated programming which is impossible to program by the manual programming. The more complicated shape and technological process the parts have, the clearer the superiority of the automatic programming is.

The automatic programming can be classified into the language-type programming (e.g., automatically programmed tools) and the conversational programming. In the language-type programming, the machining sequence, part's shape, and tools are defined in a language that can be understood by humans. The human-understandable language is then converted into a series of CNC-understandable instructions. In the conversational programming, the CNC programmer inputs the data for the part's shape interactively using a GUI (graphical user interface), selects machining sequences, and inputs the technological data for machining operations. Finally, the CNC system generates the program based on the CNC programmer's inputs. Typically, the conversational programming can be carried out by an external CAM system and a symbolic conversational system that is located either inside the CNC system or in an external computer.

2.2 Basis of the CNC Programming

2.2.1 CNC Coordinate Systems

At any time, the location of CNC machine tools is controlled by a system of XYZ coordinates called Cartesian coordinate system. This system is composed of three directional lines which are called axes, mutually intersecting at an angle of 90°. The point of intersection is known as the origin.

Almost everything that can be produced on a conventional machine tool can be produced on a CNC machine tool. A CNC machine tools movements used in producing a product have three basic control types: point-to-point control, straight-cut control, and contouring control (continuous control).

The Cartesian coordinate system was devised by the French mathematician and philosopher René Descartes. With this system, any specific point can be described in mathematical terms from one point along three perpendicular axes. This concept fits CNC machine tools perfectly. Their constructions are generally based on the three axes of motion (X, Y, Z) plus an axis of rotation (Figure 2.2). On a plain vertical CNC milling machine, the X axis is the horizontal movement (right or left) of the table, the Y axis is the table cross movement (toward or away from the column), and the Z axis is the vertical movement of the knee or the spindle. CNC systems heavily rely on the Cartesian coordinate system because the CNC programmer can locate every point on a workpiece precisely.

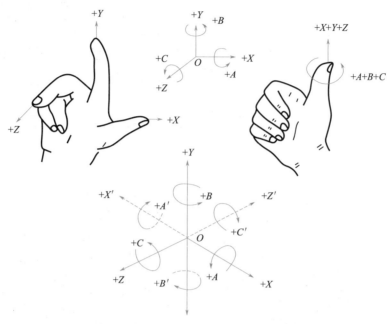

Figure 2.2　Cartesian Coordinate System

In generally, the motions of a CNC machine tool have 6 degrees of freedom. As shown in Figure 2.3, the motions can be resolved into 6 axes, namely, 3 linear axes (X, Y, and Z axis) and 3 rotational axes (A, B, and C axis).

Controllable feed and rotation motions in the workpiece machining are required on CNC machine tools, and the feed servo system can adjust the motion speed of each independent axis. The hand wheels of typical conventional machine tools are consequently redundant in a modern CNC machine tool.

A CNC lathe has at least 2 controllable or adjustable feed axes marked as X and Z (Figure 2.4). In the CNC milling machine, the main function of the workpiece clamping devices is correctly positioning the workpieces. The axes on a CNC milling machine are shown in Figure 2.5. The workpiece clamping devices should allow a workpiece change quickly, easy to approach, correctly and exactly positioned, reproducibly. For simple machining, hydraulic chuck jaws are sufficient. For milling, on all sides, the complete machining should be possible with as few reclamping as possible. The machine tool coordinate system and the workpiece coordinate system are shown in Figure 2.6 and Figure 2.7.

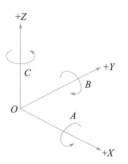

Figure 2.3　The Motions of a CNC Machine Tool

【拓展视频】

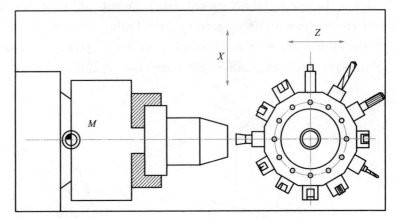

Figure 2.4　The Axes on a CNC Lathe

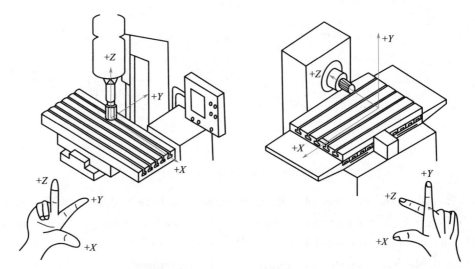

Figure 2.5　The Axes on a CNC Milling Machine

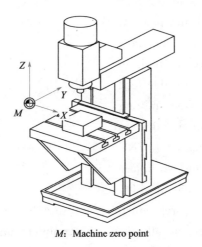 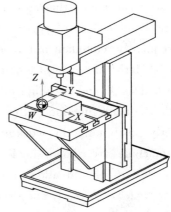

M: Machine zero point　　　　　　　　W: Workpiece zero point

Figure 2.6　The Machine Tool Coordinate System　　Figure 2.7　The Workpiece Coordinate System

Complicated milling parts with integrated automatic rotation complete machining without reclamping. Workpiece pallets, which are loaded with the next workpiece by the operator outside the work room and then automatically taken into the right machining position, are increasingly being used.

【拓展视频】

2.2.2 Coordinate Systems

1. The Absolute Coordinate System

In an absolute coordinate system, all references are made to the origin of the coordinate system. All commands of the motion are defined by the absolute coordinate referred to the origin.

2. The Incremental Coordinate System

The incremental coordinate is always used as a reference to the preceding point in a sequence of points. The disadvantage of this system is that if an error occurs, the error will be accumulated.

2.2.3 Zero Points and Reference Points

The location of zero points and reference points for a milling machine is shown in Figure 2.8.

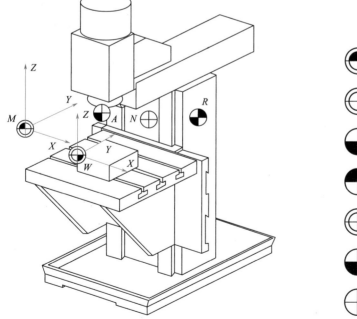
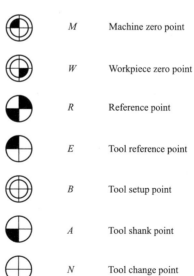

Figure 2.8 The Location of Zero Points and Reference Points for A Milling Machine

1. Machine Zero Point *M*

Each CNC machine tool works with a machine coordinate system. The machine zero point is the origin of the machine-referenced coordinate system. It is specified by the machine manufacturer, and its position cannot be changed. In general, the machine zero point *M* is located in the center of the work spindle nose for CNC lathes and above the left corner edge of the workpiece carrier for vertical CNC milling machines.

2. Workpiece Zero Point *W* (Program Zero Point)

The workpiece zero point is related to the origin position on the graphics. It is a logical reference point from which to work. The CNC programmer must decide where the workpiece zero point should be located. This decision is usually based on the convenience and easiness of programming. For example, the workpiece zero point might be located at one of the corners of the workpiece. If the workpiece is symmetrical, the workpiece zero point might be most conveniently defined at the center of the symmetry. Wherever the location is, the workpiece zero point is communicated to the machine tool operator. At the beginning of machining, the operator must move the tool under manual control to some target point on the worktable, where the tool can be easily and accurately positioned. The CNC programmer has previously referenced the target point to the machine zero point. The operator inputs the coordinate value of the workpiece zero point to the MCU.

3. Reference Point *R*

A machine tool with an incremental coordinate system needs a calibration point which also serves for controlling the movements of tool and workpiece. This calibration point is called the reference point *R*. Its location is set exactly by a limit switch on each travel axis. The coordinates of the reference point, with reference to the machine zero point, always have the same value. This value has a set adjustment in the CNC system. After switching the machine tool on, the reference point has to be approached from all axes to calibrate the incremental travel path measuring system.

4. Tool Point

If measuring tools outside of the machine tool, then the reference point on the tool, the tool holder (set up), and the tool shank are of importance, because the control must reference the geometry information (such as length and radius) of a tool to a certain point in order to apply the coordinate values from the machining program to the workpiece precisely. The tool change should be in a safe location (tool change point).

2.3 Definition of the CNC Programming

The CNC programming is where all the machining data are compiled and where the machining data are translated into a language which can be understood by the control system of the CNC machine tool. The machining data are as follows:

【拓展图文】

(1) Manufacturing methods and machining sequences, e.g., tool start up point, cutting depth, tool path, etc.

(2) Cutting conditions, e.g., spindle speed, feed rate, coolant, etc.

(3) Selection of tools.

A CNC program consists of blocks, words, and addresses (Figure 2.9 and Table 2-1).

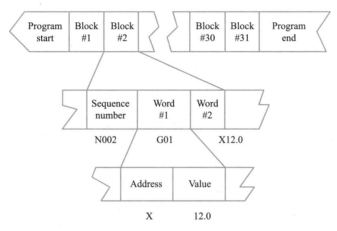

Figure 2.9 The Structure of a CNC Program

Table 2-1 Addresses

Addresses	Codes/Syntax	Functions
% MM	ISO	Subprogram
% PM	ISO	Main program
% TM	ISO	Tool compensation table
N	9001—9999999	Program number and subprogram number
N	1—8999	Block number
X	+/− 9999,99	Distance/move in mm
Y	+/− 9999,99	Distance/move in mm
Z	+/− 9999,99	Distance/move in mm
B	+/− 9999,99	Distance/move in degrees
R	+/− 9999,99	Circle radius in mm
I	+/− 9999,99	Circle center in X
J	+/− 9999,99	Circle center in Y
K	+/− 9999,99	Circle center in Z
P	0—9999	Number of subprogram calls

continued

Addresses	Codes/Syntax	Functions
F	1—5000	Feed rate in mm/rev or mm/min
S	20—99999	Spindle speed in r/min
T	0—99	Tool offset number
E	0—99	Parameter in subprograms

(1) Block: A command given to the control unit is called block.

(2) Word: A block is composed of one or more words. A word is composed of an identification letter and a series of numerals, e.g., the command for a feed rate of 200mm/min is F200.

(3) Address: The identification letter at the beginning of each word is called address. Each block (program line) contains addresses which appear in the following order:

N_ G_ X_ Y_ Z_ F_ S_ T_ M_ ;

This order should be maintained throughout every block in the program, although individual block may not necessarily contain all of these addresses.

An example of a program is as follows:

N20 G01 X20.5 F200 s1000 M03;
N21 G02 X30.0 Y40.0 I20.5 J32.0;

1. Sequence Number (N Address)

A sequence number is used to identify the block. It is always placed at the beginning of the block and can be regarded as the name of the block. The sequence number should not be consecutive. The execution sequence of the program is according to the actual sequence of the block, rather than the sequence of the number. In fact, some CNC systems do not require sequence numbers.

2. Preparatory Function (G-codes)

A preparatory function determines how the tool is moved to the programmed target. The most common G-codes and their functions (CNC lathes) are listed in Table 2-2.

Table 2-2 G-codes and their Functions (CNC Lathes)

G-codes	Functions
♯ G00	Positioning (rapid feed)
G01	Straight interpolation
G02	Circular interpolation (CW: clockwise)
G03	Circular interpolation (CCW: counter-clockwise)
G04	Dwell

continued

G-codes	Functions
G20	Data input (inch)
#G21	Data input (mm)
#G22	Stored distance limit is effective (spindle interference check ON)
G23	Stored distance limit is ineffective (spindle interference check OFF)
G27	Machine reference point return check
G28	Automatic reference point return
G29	Return from reference point
G30	The second reference point return
#G32	Thread process
G40	Cancel of compensation
G41	Tool compensation (left)
G42	Tool compensation (right)
G50	Set the coordinate system/Limit the spindle speed
G70	Compound repeat cycle (finishing cycle)
G71	Compound repeat cycle (stock removal in turning)
G72	Compound repeat cycle (stock removal in facing)
G73	Compound repeat cycle (pattern repeating cycle)
G74	Compound repeat cycle (peck drilling in Z direction)
G75	Compound repeat cycle (grooving in X direction)
G76	Compound repeat cycle (thread process cycle)
G90	Absolute programming
G91	Incremental programming
G92	Fixed cycle (thread process cycle)
G94	Fixed cycle (facing process cycle)
G96 #G97	Control the circumference speed uniformly/Constant surface speed control (mm/min) Cancel the uniform control of circumference speed/Cancel the constant surface speed control r/min
G98	Designate the feedrate per minute (mm/min)
#G99	Designate the feedrate per the rotation of principal spindle (mm/r)

Note: 1. # mark instruction is the modal indication of initial condition which is immediately available when power is supplied.

2. In general, G-codes are used in CNC lathes, and it is possible to select the special G codes according to the setting of parameters.

3. Parameters for Circular Interpolation (I/J/K Addresses)

These parameters specify the distance measured from the start point of the arc to the center. Numerals following I, J, and K are the X, Y, and Z components of the distance, respectively.

4. Spindle Function (S Address)

The spindle speed is commanded under an S address, and its unit is always in revolution per minute. It can be calculated by the following formula:

$$\text{Spindle Speed} = \frac{\text{Surface Cutting Speed} \times 1000}{\pi \times \text{Tool Diameter}} \quad (2.1)$$

5. Feed Function (F Address)

The feed is programmed under an F address except for rapid traverse. The unit may be in mm per minute (in the case of milling machine) or in mm per revolution (in the case of turning machine). The unit of the feed rate has to be defined at the beginning of the program. The feed rate can be calculated by the following formula:

$$\text{Feed Rate} = \text{Chip Load} \times \text{No. of Tooth} \times \text{Spindle Speed} \quad (2.2)$$

Where, Chip load—Feed per tooth.

6. Miscellaneous Function (M-codes)

The miscellaneous function is programmed to control the operation of CNC machine tools rather than the coordinate movements. M-codes that are especially useful in many programming applications and their functions are shown in Table 2-3.

Table 2-3 M-codes and their Functions

M-codes	Functions
M00	Program stop
M02	End of a program
M03	Spindle start (CW: clockwise)
M04	Spindle start (CCW: counter-clockwise)
M05	Spindle stop
M06	Manual/automatic tool change with the automatic travel to a fixed machine-specific change position
M07	Internal cooling lubricant supply ON
M08	External cooling lubricant supply ON
M09	Cooling lubricant supply OFF
M10	Chuck-clamping
M11	Chuck-unclamping
M12	Tailstock spindle out
M13	Tailstock spindle in
M19	Spindle stop with a defined final position

Continued

M-codes	Functions
M30	Program end with a reset of the CNC system to a ready condition
M66	Manual tool change in the position last moved to
M67	Tool data activation without a tool change
M98	Transfer to a subprogram
M99	End of a subprogram

2.4 CNC Programming Examples

【拓展视频】

There are many codes included in a program. G-codes perform preparatory functions, and M-codes perform auxiliary functions. They are the base of CNC programs. ISO (International Organization for Standardization) has worked out the standards of G-codes and M-codes. Because new CNC systems and CNC machine tools have been emerging, a lot of functions in many systems surpass ISO standards. Their codes are abundant, and their formats are flexible, which are not restrained by ISO standards. In addition, even with the same function, the codes and the formats are different among CNC systems made by different companies. The codes and formats are also different between new and old CNC systems made in the same company, but G-codes and M-codes in most CNC systems follow the ISO standards.

Usually, several tools are used for machining one workpiece. The tools have different lengths (Figure 2.10). It is very troublesome to change the program in accordance with tools.

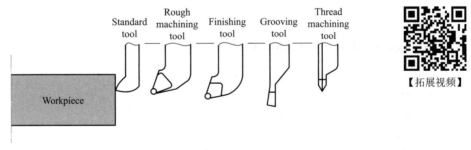

Figure 2.10 Different Lengths of Tools

【拓展视频】

Therefore, the length of each tool used should be measured in advance. By setting the difference between the length of the standard tool and the length of each tool in the CNC, machining can be performed without altering the program even when the tool is changed. This function is called tool length compensation.

2.4.1 Turning Programming of MAHO GR350C

The front panel of MAHO GR350C is shown in Figure 2.11.

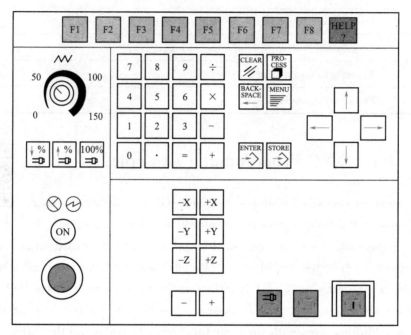

Figure 2.11 The Front Panel of MAHO GR350C

【拓展视频】

One program example（MAHO GR350C）：

% PM1234567　　　　　　　　（% PM+ N　N-Program number）
N1234567　　　　　　　　　（Body）
N1 G18;
N2 G52;
...
N2000 M30;　　　　　　　　（End）

1. G10：Axial Volumetric Cutting Cycle

Format：G10 X_(U_)Z_(W_)I_K__C_(F_)

Where，X—Start S：X (absolute dimensioning).

U—Start S：X (incremental dimensioning).

Z—Start S：Z (absolute dimensioning).

W—Start S：Z (incremental dimensioning).

I—X Finishing allowance (G12).

K—Z Finishing allowance (G12).

C—Cutting depth per time.

F—Feed speed，mm/r.

G10 example (Figure 2.12):

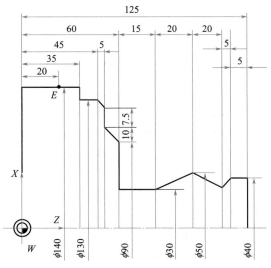

Figure 2.12 G10 Example

```
% PM241001
N241001
N1 G54;                                    (Definition of the programming zero point)
N2 G99 X140 Z127;                          (Blank definition)
N3 G96 S100 D2500 T1013 M4;                (Constant surface speed selection, Linear ve-
                                           locity 100m/min, Maximum spindle speed 2500 r/
                                           min)
N4 G10 X145 Z130 I0.5 K0.5 C2.5 F0.5;      (Definition, Finishing allowance X0.5 Z0.5, cutting
                                           depth/P2.5, Start 145,130)
N8001 G01 X40 Z125;                        (Definition of contour Start)
N5 G01 W-5;
N6 G01 X30 W-5;
N7 G01 X50 Z95;
N8 G01 X30 Z75;
N9 G01 Z60;
N10 G01 X90;
N11 G01 U20 Z50;
N12 G01 U15;
N13 G01 X130 Z45;
N14 G01 Z35;
N15 G01 X140;
N16 G01 Z28;                               (Definition of contour End)
N17 G13 N1= 8001 N2= 16;                   (Program calls: Roughing the outside contour)
N18 G00 X200 Z145;                         (Approaching the tool change position)
N19 T2 023;                                (Selection of the tool)
N20 G12 X145 Z127 S200;                    (Finishing the outside contour: Definition)
N21 G13 N1= 8001 N2= 16;                   (Program calls)
```

N22 G00 X300 Z350;   (Tool retracting)
N23 M30;

2. G11：Radial Volumetric Cutting Cycle（Figure 2.13）

Format：G11 x_(U_)Z_(W_)I_K__C_(F_)

Where，X—Start S：X（absolute dimensioning）.

U—Start S：X（incremental dimensioning）.

Z—Start S：Z（absolute dimensioning）.

W—Start S：Z（incremental dimensioning）.

I—X Finishing allowance (G12).

K—Z Finishing allowance (G12).

C—Cutting depth per time.

F—Feed speed，mm/r.

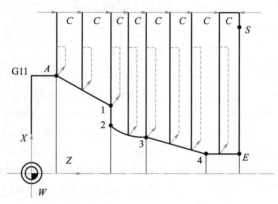

Figure 2.13 G11 Function

G11 example（Figure 2.14）：

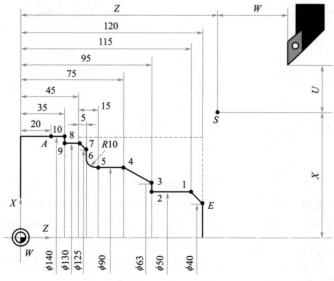

Figure 2.14 G11 Example

```
% PM241002
N241002
N1 G54;                              (Workpiece coordinate system)
N2 G99 X140 Z127;                    (Blank definition)
N3 G96 S100 D2500 T1013 M4;          (Constant surface speed selection, Linear ve-
                                      locity 100m/min, Maximum spindle speed 2500 r/
                                      min)
N4 G00 X200 Z145;                    (Setting at the start point)
N5 G11 X145 Z120 I0.5 K0.5 C2.5 F0.5; (Definition, Finishing allowance X0.5 Z0.5,
                                      Cutting depth/P2.5, Start 145,120)
N6 G01 Z25;                          (Definition of contour Start)
N7 G41 X140;                         (Tool compensation, Left)
N8 G01 Z35;
N9 G01 X130;
N10 G01 Z45 F0.05;
N11 G01 X125 Z50;
N12 G01 X110;
N13 G03 X90 Z60 R10;
N14 G01 Z75;
N15 G01 X63 Z95 F0.1;
N16 G01 X50;
N17 G01 Z115;
N18 G01 X40 Z120;                    (Definition of contour End)
N19 G40;                             (Cancel of tool compensation)
N20 G13 N1=6 N2=19;                  (Program calls: Roughing the outside contour)
N21 G00 X200 Z145;                   (Approaching the tool change position)
N22 T2 023;                          (Selection of the tools)
N23 G12 X145 Z120 S200;              (Finishing the outside contour: Definition)
N24 G13 N1=6 N2=19;                  (Program calls)
N25 G00 X300 Z350;
N26 M30;
```

3. G12: Contour Finishing Cycle Example

Format: G12 X_(U_)Z_(W_)(F_)

Where, X—Start S: X (absolute dimensioning).
U—Start S: X (incremental dimensioning).
Z—Start S: Z (absolute dimensioning).
W—Start S: Z (incremental dimensioning).

4. G13: Volumetric Cutting Cycle Call (Execution)

Format: G13 N1=_ N2=_

Where, N1—Definition contour, program number: start.

N2—Definition contour, program number: end.

5. G96: Constant Surface Speed Selection, G97: Constant Surface Speed Cancel

Format: G96 S_ F_ D_

Where, S—Linear velocity (m/min).

F—Feed speed.

D—Maximum spindle speed (r/min).

Format: G97 S_

Where, S—Maximum spindle speed (r/min).

6. G32: Thread Cutting Cycle

Format: G32 X_(U_)Z_(W_)C_(D_)(A_)(J_)(B_)F_

Where, X—Bottom diameter of thread.

Z—End point of Z axis.

U—Depth of thread, +U internal thread, −U outside thread.

W—Length of thread.

C—1st cut depth, if C=U, cut only one; If C<U, cut repeatedly.

D—Cutting depth of finishing.

A—Half of the angle between threads.

J—End diameter of taper cut.

B—Taper, 1 : P.

F—Thread pitch.

G32 parameters are shown in Figure 2.15.

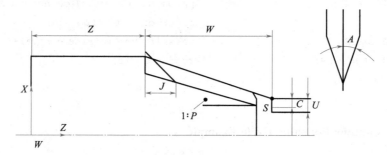

Figure 2.15 G32 Parameters

2.4.2 Turning Programming of FANUC 0i MATE TB

G-codes and their functions of FANUC 0i MATE TB are shown in Table 2-4.

Table 2-4 G-codes and their Functions of FANUC 0i MATE TB

G-codes	Functions	G-codes	Functions
G00	Positioning (rapid traverse)	G57	Workpiece coordinate system 4 selection
G01	Linear interpolation (cutting feed)	G58	Workpiece coordinate system 5 selection
G02	Circular interpolation (CW)	G59	Workpiece coordinate system 6 selection
G03	Circular interpolation (CCW)	G65	Macro call
G04	Dwell	G66	Macro modal call
G10	Programmable data input	G67	Macro modal call cancel
G11	Programmable data input cancel	G70	Finish turing cycle
G12	Input in inch	G71	Stock removal in turning
G13	Input in mm	G72	Stock removal in facing
G27	Reference position return check	G73	Pattern repeat
G28	Return to reference position	G74	End face peck drilling
G31	Skip function	G75	Outer diameter/Internal diameter drilling
G32	Thread cutting	G76	Multiple threading cycle
G40	Tool nose radius compensation cancel	G90	Absolute programming
G41	Tool nose radius compensation (left)	G91	Incremental programming
G42	Tool nose radius compensation (right)	G92	Thread cutting cycle
G50	Coordinate system set/Maximum spindle speed set	G94	End face turning cycle
G52	Local coordinate system set	G96	Constant surface speed
G53	Machine coordinate system set	G97	Constant surface speed cancel
G54	Workpiece coordinate system 1 selection	G98	Feed speed in mm/min
G55	Workpiece coordinate system 2 selection	G99	Feed speed in mm/r
G56	Workpiece coordinate system 3 selection		

1. G01: Linear Interpolation (Cutting Feed)

Format: G01 X(U)_Z(W)_F_

G01 example (Figure 2.16):

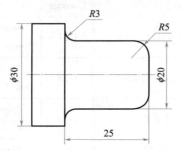

Figure 2.16　G01 Example

O2421
N10 T0101;
N20 G0 X0 Z1. S500 M03;   (Absolute coordinate)
N30 G1Z0. F0.2;
N40 G1 X20. R-5. ;
N50 G1 Z-25. R3. ;
N60 G1 X30.5;
N70 G28 X120. Z100. ;
N80 M30;

2. G04: Dwell

Format: G04 X(U)_ or G04 P_

Where, X—Specify a time (decimal point permitted).

U—Specify a time (decimal point permitted).

P—Specify a time (decimal point not permitted).

3. G28: Return to Reference Position

Format: G28 X(U)_Z(W)

Positioning to the intermediate or reference positions are performed at the rapid traverse rate of each axis. Therefore, for safety, the tool nose radius compensation and the tool offset should be cancelled before executing this command.

4. G32: Thread Cutting

Format: G32 X(U)_Z(W)_F_

In general, thread cutting is repeated along the same tool path in rough cutting through finish cutting for a screw. Since thread cutting starts when the position coder mounted on the spindle outputs a one-turn signal, threading is started at a fixed point and the tool path on the workpiece is unchanged for repeated thread cutting. Note that the spindle speed must remain constant from rough cutting to finish cutting. If not, incorrect thread lead will occur.

G32 example (Figure 2.17):

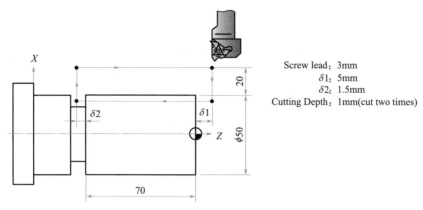

Screw lead: 3mm
$\delta 1$: 5mm
$\delta 2$: 1.5mm
Cutting Depth: 1mm(cut two times)

Figure 2.17 G32 Example

O2422

N10 G50;

N20 G97 S800 M03;

N30 G00 X90.0 Z5.0 T0101 M8;

N40 X48.0;

N50 G32 Z-71.5 F3.0;

N60 G00 X90.0;

N70 Z5.0;

N80 X46.0;

N90 G32 Z-71.5;

N100 G00 X90.0;

N110 Z5.0

N120 X150.0 Z150.0;

N130 M30;

5. G50: Coordinate System Set or Spindle Speed Set

G50 is used for setting the coordinate system or the maximum spindle speed.

6. G90: Absolute Programming

G91: Incremental Programming

Format: G90 X(U)_ Z(W)_ F_

There are two ways to command travels of the tool: the absolute command and the incremental command. In the absolute command, the coordinate value of the end position is programmed. In the incremental command, the movement distance of the position itself is programmed.

G90 and G91 are used for the absolute programming and the incremental programming, respectively.

G90 example 1 (Figure 2.18):

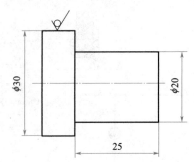

Figure 2.18　G90 Example 1

O2423
N10 T0101;
N20 G0 X31. Z1. S800 M03;   (Start point)
N30 G90 X26. Z-24. 9 F0. 3;   (X 2mm,machining allowance 0.1mm for finish machining)
N40 X22. ;
N50 X20. 5;   (X 0.25mm)
N70 X20. Z-25. F0. 2 S1200;   (Finish machining)
N80 G28 X100. Z100. ;
N90 M30;

G90 example 2 (Figure 2.19):

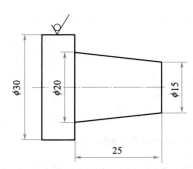

Figure 2.19　G90 Example 2

O2424
N10 T0101;   (Tool)
N20 G0 X32. Z0. 5 S500 M3;
N30 G90 X26. Z-25. R-2. 5 F0. 15; (Rough machining)
N40 X22. ;
N50 X20. 5;   (Machining allowance 0.5mm for finish machining)
N60 G0 Z0 S800 M3;
N70 G90 X20. Z-25. R-2. 5 F0. 1;
N80 G28 X100. Z100. ;

N90 M5;
N100 M2;

7. G92: Thread Cutting Cycle

Format: G92 X(U)_Z(W)_F_

The range of thread leads, limitation of spindle speed and other parameters, are the same as in G32 (thread cutting). Thread chamfering can be performed in this thread cutting cycle. A signal from the machine tool initiates thread chamfering. The chamfering distance is specified in a range from 0.1L to 12.7L in 0.1L increments by the No. 5130 parameter. In the above expression, L is the thread lead.

G92 example 1 (Figure 2.20):

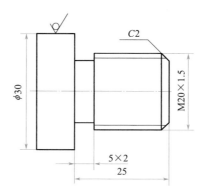

Figure 2.20 G92 Example 1

```
O2425
N110 T0303;
N120 G0 X28. Z5. S350 M3;    (Start point)
N130 G92 X19.4 Z-23. F1.5;   (Thread cutting)
N140 X19.;                   (Cycle)
N150 X18.6;
N160 X18.2;
N170 X18.;
N180 X17.9;
N190 X17.8;
...
```

G92 example 2 ($P = 1.5$, Figure 2.21):

Format: G92 X(U)_Z(W)_R_F_

Where, F—Screw lead (L) is specified.

R—Difference between the end radius of the taper thread and the starting radius.

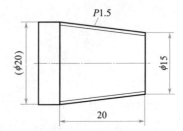

Figure 2.21   G92 Taper Example 2

```
O2426
N10 T0101;
N20 G0 X25. Z5. S300 M3;
N30 G92 X19.6 Z-20. R-2.5 F1.5;
N40 X19.4;
N50 X19.;
```

8. G94：End Face Turning Cycle

Format：G94 X(U)_Z(W)_F_

The parameters of the end face turning cycle are shown in Figure 2.22.

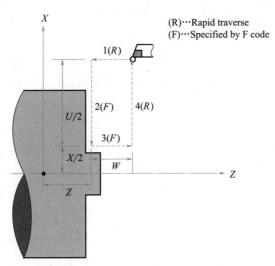

Figure 2.22   The Parameters of the End Face Turning Cycle

In the incremental programming, the sign of numbers following addresses U and W depends on the direction of paths 1 and 2. That is, if the direction of the path is in the negative direction of the Z axis, the value of W is negative.

In the single block mode, operations 1, 2, 3, and 4 are performed by pressing the cycle start button once.

G94 can be used as taper face cutting cycle.

G94 example (Figure 2.23)：

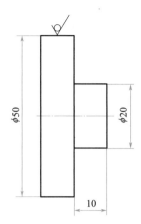

Figure 2.23 G94 Example

```
O2427
N10 T0101;
N20 G0 X52. Z1. S500 M03;
N30 G94 X20.2 Z-2. F0.2;     (Rough machining, Z 2)
N40 Z-4.;
N50 Z-6.;
N60 Z-8.;
N70 Z-9.8;
N80 X20. Z-10. S1200;        (Finish machining)
N90 G28 X100. Z100.;
N100 M30;
```

9. G70: Finish Turning Cycle

Format: G70 P(ns)_Q(nf)_

Where, (ns) —Sequence number of the first block for the program of finishing shape.

(nf) —Sequence number of the last block for the program of finishing shape.

After rough cutting by G71, G72, or G73, the following command permits finish turning cycle.

NOTE:

(1) F, S, and T functions specified in the block G71, G72, and G73 are not effective, but those specified between sequence numbers "ns" and "nf" are effective in G70.

(2) When the cycle machining by G70 is terminated, the tool is returned to the start point, and the next block is read.

(3) In blocks between "ns" and "nf" referred in G70, G71, G72, and G73, the subprogram cannot be called.

10. G71: External Rough Turning Cycle

Format: G71 U(Δd)_R(e)_
G71 P(ns)_Q(nf)_U(Δu)_W(Δw)_F_s_T_

Where, Δd—Depth of cut (radius designation). Designate without signs. The cutting direction depends on the direction of AA′. This designation is modal and is not changed until the other value is designated. Also, this value can be specified by the parameter (No. 5132), and the parameter is changed by the program command.

e—Escaping amount. This designation is modal and is not changed until the other value is designated. Also, this value can be specified by the parameter (No. 5133), and the parameter is changed by the program command.

ns—Sequence number of the first block for the program of finishing shape.

nf—Sequence number of the last block for the program of finishing shape.

Δu—Distance and direction of the finishing allowance in the X direction (diameter/radius designation).

Δw—Distance and direction of the finishing allowance in the Z direction.

F, S, T—Any F, S, or T function contained in blocks "ns" and "nf" in the cycle is ignored. Any F, S, or T function in the G71 block is effective.

If a finished shape of A to A′ to B is given by a program as in figure 2.24, the specified area is removed by Δd, with finishing allowance Δu/2 and Δw left.

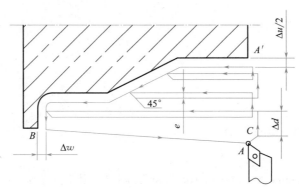

Figure 2.24 G71 Parameters

G71 example (Figure 2.25):

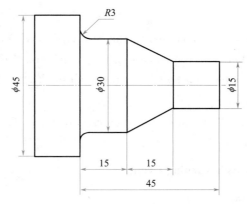

Figure 2.25 G71 Example

O2428
N10 T0101;
N20 G0 X46. Z0. 5 S500 M03;
N30 G71 U2. R0. 5;   (Depth of cut 2mm)
N40 G71 P50 Q110 U0. 3 W0. 1 F0. 3;   (Machining allowance of finish machining X 0. 3mm, Z 0. 1mm; Feed speed of rough machining 0. 3mm/r)
N50 G1 X15. ;
N60 G1 Z0. F0. 15 S1500;   (Feed speed of finish machining 0. 15mm/r; Spindle speed of finish machining 1500r/min)
N70 Z-15. ;
N80 X30. Z-30. ;
N90 Z-42. ;
N100 G2 X36. Z-45. R3. ;
N110 G1 X46. ;
N120 G70 P50 Q100;   (Finish machining cycle)
N130 G28 X100. Z100. ;
N140 M5;
N150 M30;

11. G72: Rough Facing Cycle

Format: G72 W(d)_R(e)_
G72 P(ns)_Q(nf)_U(u)_W(w)_F_S_T_

As shown in Figure 2. 26, this cycle is as same as G71 except that cutting is made by an operation parallel to the X axis.

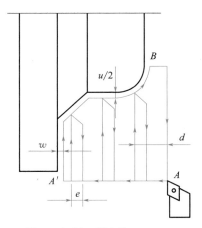

Figure 2. 26 G72 Parameters

G72 example (Figure 2. 27):

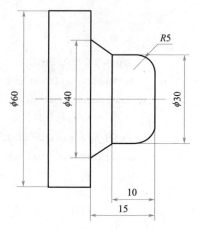

Figure 2.27 G72 Example

O2429
N10 T0101;
N20 G0 X61.Z0.5 S500 M03;
N30 G72 W2.R0.5;
N40 G72 P50 Q100 U0.1 W0.3 F0.25;
N50 G0 Z-15.;
N60 G1 X40.F0.15 S800;
N70 X30.Z-10.;
N80 Z-5.;
N90 G2 X20.Z0 R5.;
N100 G0 Z0.5;
N110 G70 P60 Q110;
N120 G28 X100.Z100.;
N130 M30;

12. G73：Pattern Repeating Cycle

Format：G73 U(Δi)_W(Δk)_R(Δd)
G73 P(ns)_Q(nf)_U(Δu)_W(Δw)_F_s_T_

This function permits cutting a fixed pattern repeatedly, with a pattern being displaced bit by bit. By this cutting cycle, it is possible to efficiently cut a workpiece whose rough shape has already been made by a rough machining, forging, or casting.

G73 example (Figure 2.28)：

O24210
N10 T0101;
N20 G0 X110.Z10.S800 M3;
N30 G73 U5.W3.R3.;
N40 G73 P50 Q110 U0.4 W0.1 F0.3;
N50 G0 X50.Z1.S1000;

N60 G1 Z-10. F0.15;
N70 X60. Z-15.;
N80 Z-25.;
N90 G2 X80. Z-35. R10.;
N100 G1 X90. Z-40.;
N110 G0 X110. Z10.;
N120 G70 P50 Q110;
N130 G28 X100. Z150.;
N140 M30;

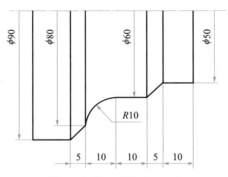

Figure 2.28 G73 Example

13. G74: End Face Peck Drilling

Format: G74 R(e)_
G74 X(U)_Z(W)_P(Δi)_Q(Δk)_R(Δd)_F_

Where, e—Return amount. This designation is modal and is not changed until the other value is designated. Also, this value can be specified by the parameter No. 5139, and the parameter is changed by the program command.

X—X component of point B.

U—Incremental amount from A to B.

Z—Z component of point C.

W—Increment amount from A to C.

Δi—Movement amount in X direction (without sign), P1000 is 1mm.

Δk—Depth of cut in Z direction (without sign).

Δd—Relief amount of the tool at the cutting bottom. The sign of Δd is always plus (+). However, if address X (U) and Δi are omitted, the relief direction can be specified by the desired sign.

F—Feed rate.

G74 example (Figure 2.29):

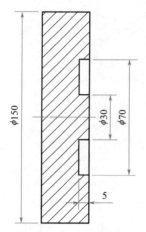

Figure 2.29　G74 Example

```
O2430
N10 T0606;                    (Width of tool 4mm)
N20 S300 M3;
N30 G0 X30. Z2. ;
N40 G74 R1. ;
N50 G74 X62. Z-5. P3500 Q3000 F0. 1;
N60 G0 X200. Z50. M5;
N70 M30;
```

Application of G-codes example (Figure 2.30):

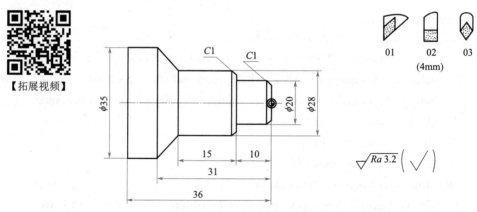

Figure 2.30　Application of G-codes Example

```
O2431                         (O Program number)
N10 T0101;
N20 S1000 M03;
N30 G00 X40. Z2. ;
N40 G71 U1. 5 R1. ;           (U-Depth of cut, R-Escaping amount)
N50 G71 P60. Q150. U0. 5 W0. 2 F0. 3;   (P-Sequence number of the first block, Q-Sequence
                                         number of the last block, U-X machining allowance,
```

			(W-Z machining allowance)
N60	G01	X18. Z0.;	(Start point)
N80		X20. Z-1.;	(Chamfer 1×45°)
N90	G01	Z-10.;	(φ20mm)
N100		X26.;	
N110		X28. Z-11.;	(Chamfer 1×45°)
N120		Z-25.;	
N140		X35. Z-31.;	
N150		Z-36.;	(End point)
N160	G70	P60 Q150 S1100 F0.05;	(N60—N150)
N170	G00	X50. Z60.;	
N180	T0202;		
N190	S200 M03;		(Cut off by right edge)
N210	G00	X37. Z-40.;	(Length 36mm)
N220	G01	X0. F0.03;	
N230	G00	X50.;	
N240	G00	Z0.;	
N250	M05;		
N260	M02;		(End)

2.4.3 Milling Programming of MAHO 600C

MAHO 600C have four axes (B, X, Y, Z). The interface of MAHO 600C is shown in Figure 2.31. G-codes of MAHO 600C and their functions are shown in Table 2-5.

【拓展视频】

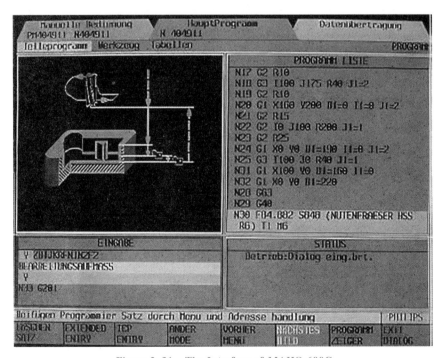

Figure 2.31　The Interface of MAHO 600C

Table 2-5　G-codes of MAHO 600C and Their Functions

G-codes	Functions
G00*	Rapid positioning
G01	Linear interpolation
G02	Clockwise circular interpolation
G03	Counter clockwise circular interpolation
G04**	Dwell 0s, or 1—983s
G17*	Plane selection *XOY*
G18	Plane selection *XOZ*
G19	Plane selection *YOZ*
G22**	Subprogram selection
G23**	Main program selection
G25	Manual feed correction (F-over) ON
G26	Manual feed correction (F-over) OFF
G27	Feed with rounding
G28	Feed with precision stop
G40*	Cancel of G41, G42, G43, G44
G41	Tool radius compensation (left)
G42	Tool radius compensation (right)
G43	Tool length compensation (positive)
G44	Tool length compensation (negative)
G51*	Cancel of G52
G52	Selection of the zero point (Reset Axis)
G53*	Cancel of G54, G55, G56, G57, G58, G59
G54	Stored Zero point 1 Selection
G55	Stored Zero point 2 Selection
G56	Stored Zero point 3 Selection
G57	Stored Zero point 4 Selection
G58	Stored Zero point 5 Selection
G59	Stored Zero point 6 Selection
G63*	Cancel of G64

Continued

G-codes	Functions
G64	Activation of the geometry processor
G70	Dimensions in inches (inch dimension system)
G71*	Dimensions in mm (mm dimension system)
G72*	Cancel of G73
G73	Mirror image machining/Scale up and scale down (in combination with the word A4=)
G74**	Travel to an absolute position referred directly to the reference point R of the machine in rapid traverse
G77**	Cycle selection with point circle definition
G78**	Point definition
G79**	Cycle selection
G81	Drilling cycle
G83	Drilling cycle with chip clearance/chip breaking (deep hole drilling cycle)
G84	Tapping cycle
G85	Reaming cycle
G86	Boring cycle
G87	Rectangular recess milling cycle
G88	Slot milling cycle
G89	Circular recess milling cycle
G90*	Absolute programming (reference dimensions)
G91	Incremental programming (chain dimensions)
G92	Programmed incremental zero point shift. Rotation of the coordinate system (in conjunction with the word B4=)
G93	Programmed absolute zero point shift. Rotation of the coordinate system (in conjunction with the word B4=)
G94*	Feed speed in mm per minute (G71) or in inches per minute (G70)
G95	Rotational feed in mm per revolution (G71) or in inches per revolution (G70)
G98	Definition of window for graphic test runs
G99	Definition of workpiece for graphic test runs

Note: 1. * Functions are active in the read condition of the CNC control system.
      2. ** Active only in sentence.

Milling (Thickness 20mm) example 1 (Figure 2.32):

【拓展视频】

【拓展视频】

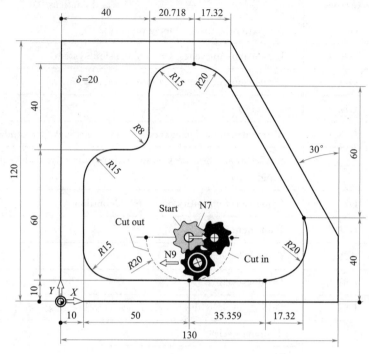

Figure 2.32 Milling (Thickness 20mm) Example 1

```
% PM24301
N24301
```

N1	G17 S900 T31 M66;	(Tool diameter 10mm)
N2	G54;	(Selection of stored zero point)
N3	G98 X-10 Y-10 Z-20 I150 J140 K30;	(Display window)
N4	G99 X0 Y0 Z-20 I130 J120 K20;	(Workpiece definition)
N5	G00 X60 Y30 Z8 M3;	
N6	G01 Z-21 F50;	(Down)
N7	G43 X80 F100;	(Tool radius compensation up to X80)
N8	G42;	(Tool compensation right)
N9	G02 X60 Y10 R20;	(Or I60 J30)
N10	G01 X25;	
N11	G02 X10 Y25 R15;	(Or I25 J25)
N12	G01 Y55;	
N13	G02 X25 Y70 R15;	(Or I25 J55)
N14	G01 X32;	
N15	G03 X40 Y78 R8;	(Or I32 J78)
N16	G01 Y95;	
N17	G02 X55 Y110 R15;	(Or I55 J95)

```
N18 G01 X60.718;
N19 G02 X78.039 Y100 R20;        (Or I60.718 J90)
N20 G01 X112.679 Y40;
N21 G02 X95.359 Y10 R20;         (Or I90.359 J30)
N22 G01 X60;
N23 G02 X40 Y30 R20;             (Or I60 J30)
N24 G40;                         (Cancel the compensation)
N25 G00 Z50 M5;                  (Approaching the tool change position)
N26 G53;                         (Cancel of G54)
N27 T0 M66;                      (Manual tool change)
N28 M30;
```

1. G40/G41/G42: Tool Compensation

In CNC machining, if the tool axis is moving along the programmed path, the dimension of the workpiece obtained will be incorrect since the tool diameter has not to be taken into account.

Modern CNC systems are capable of doing tool compensation. What the system requires are the programmed path, the tool diameter, and the tool position with reference to the contour. Normally, the tool diameter is not included in the program. It has to be input into the CNC system in the tool setting process.

G40 is used to cancel the compensation calculation.

If the tool is on the left of the contour, G41 will be used. The tool is positioned on the left-hand side of the workpiece, as seen from behind the tool, following the direction of movement.

If the tool is on the right of the contour, G42 will be used. The tool is positioned on the right-hand side of the workpiece, as seen from behind the tool, following the direction of movement.

2. G14: Repeat Section Selection

    Format: G14 N1=_ N2=_ J_

Where, N1—Program number (start).

       N2—Program number (end).

       J—Repeat times (default once).

3. G92/G93: Zero Point Shift (Figure 2.33)

    Format: G92 X_Y_Z_B1=_L1=_B4=_
            G93 X_Y_Z_B2=_L2=_B4=_

Where, X, Y, Z—Displacement.

       B1, L1—Polar (G92).

       B2, L2—Polar (G93).

       B4—The angle between two axes.

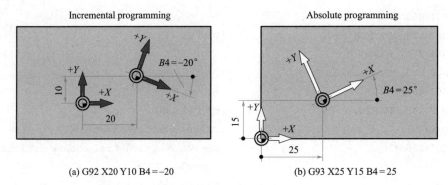

Figure 2.33　G92 and G93

Milling (Thickness 20mm) example 2 (Figure 2.34):

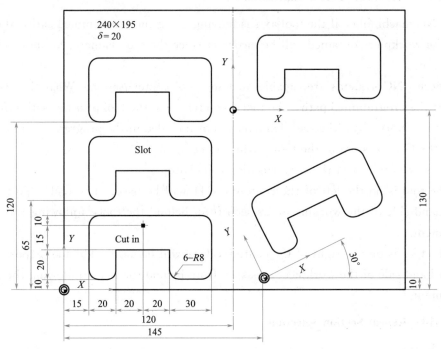

Figure 2.34　Milling (Thickness 20mm) Example 2

```
% PM24302
N24302
N1 G17 S400 T31 M66;            (Tool diameter 10mm)
N2 G54;
N3 G98 X-10 Y-10 Z-10 I260 J215 K30;
N4 G99 X0 Y0 Z-20 I240 J195 K20;   (Workpiece definition)
N5 G00 X55 Y45 Z2 M13;
N6 G1 Z-21 F50;                 (Down)
N7 G43 Y55 F100;                (Tool length compensation up to Y55)
N8 G42;                         (Tool radius compensation right)
```

```
N9  G01 X97;                    (Milling the contour)
N10 G02 X105 Y47 R8;
N11 G01 Y18;
N12 G02 X97 Y10 R8;
N13 G01 X83;
N14 G02 X75 Y18 R8;
N15 G01 Y30;
N16 G01 X35;
N17 G01 Y18;
N18 G02 X27 Y10 R8;
N19 G01 X23;
N20 G02 X15 Y18 R8;
N21 G01 Y47;
N22 G02 X23 Y55 R8;
N23 G01 X55;                    (Milling the contour End)
N24 G00 Z50;
N25 G40;                        (Cancel tool compensation)
N26 G92 Y55;                    (Programmed incremental zero point shift)
N27 G14 N1= 5 N2= 26 J2;        (Repeat twice)
N28 G93 X120 Y130;              (Programmed absolute zero point shift)
N29 G14 N1= 5 N2= 25;           (Repeat once)
N30 G93 X145 Y10 B4= 30;        (Programmed absolute zero point shift)
N31 G14 N1= 5 N2= 25;           (Repeat once)
N32 G00 Z50 M5;                 (Approaching the tool change position)
N33 G53;                        (Cancel of G54)
N34 M30;
```

4. G72/G73: Mirror Image, Scale up and Scale down

G72 is used for cancelling G73.

G73: X-1(Y-1)(Z-1), it is used for mirror image machining (Figure 2.35).

G73: A4= , it is used for scaling up and scaling down (in combination with the word A4=).

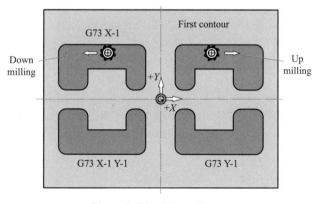

Figure 2.35 Mirror Image

Mirror image example (Figure 2.36):

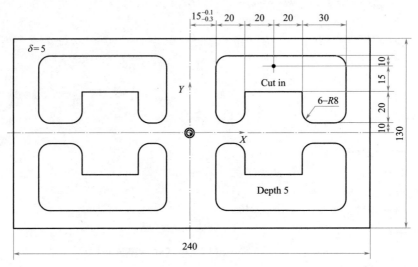

Figure 2.36　Mirror Image Example

```
% PM24303
N24303                              (Workpiece 240mm×130mm×5mm)
N1 G17 S400 T31 M66;                (Tool diameter 10mm)
N2 G54;
N3 G98 X-130 Y-75 Z-10 I260 J150 K20;
N4 G99 X-120 Y-65 Z-5 I240 J130 K5; (Workpiece definition)
N5 G0 X55 Y45 Z2 M13;               (G00)
N6 G1 Z-6 F50;                      (Down)
N7 G43 Y55 F100;                    (Tool length compensation up to Y55)
N8 G42;                             (Tool radius compensation right)
N9 G1 X97;                          (Milling the contour)
N10 G2 X105 Y47 R8;
N11 G1 Y18;
N12 G2 X97 Y10 R8;
N13 G1 X83;
N14 G2 X75 Y18 R8;
N15 G1 Y30;
N16 G1 X35;
N17 G1 Y18;
N18 G2 X27 Y10 R8;
N19 G1 X22.8;
N20 G2 X14.8 Y18 R8;
N21 G1 Y47;
N22 G2 X22.8 Y55 R8;
```

N23 G1 X55;   (Milling the contour End)
N24 G0 Z50;
N25 G40;   (Cancel tool compensation)
N26 G73 X-1;   (Mirror Y)
N27 G14 N1= 5 N2= 25;   (Repeat once)
N28 G72;
N29 G73 X-1 Y-1;   (Mirror origin)
N30 G14 N1= 5 N2= 25;   (Repeat once)
N31 G72;
N32 G73 Y-1;   (Mirror X)
N33 G14 N1= 5 N2= 25;   (Repeat once)
N34 G72;   (Cancel of G73)
N35 G0 Z50 M5;   (Approaching the tool change position)
N36 G53;
N37 M30;

2.4.4 Milling Programming of FANUC 0MC

G-codes of FANUC 0MC and their functions are shown in Table 2-6.

【拓展视频】

Table 2-6 G-codes of FANUC 0MC and their Functions

G-codes	Functions	G-codes	Functions
G00	Rapid positioning	G21	Input in mm
G01	Linear interpolation	G22	Stored stroke check function on
G02	Circular arc interpolation/Helical interpolation	G23	Stored stroke check function off
G03	Circular arc interpolation/Helical interpolation	G24	Mirror image
G04	Dwell	G25	Mirror image cancel
G09	Exact stop	G27	Reference position return check
G10	Programmable data input	G28	Return to reference position
G11	Programmable data input cancel	G29	Return from reference position
G15	Polar coordinates command cancel	G30	Return to 2nd, 3rd, and 4th reference position
G16	Polar coordinates command	G31	Skip function
G17	XOY plane selection	G33	Thread cutting
G18	XOZ plane selection	G39	Corner offset circular interpolation
G19	YOZ plane selection	G40	Tool radius compensation cancel/Three dimensional compensation cancel
G20	Input in inch	G41	Tool radius compensation (left) /Three dimensional compensation

Continued

G-codes	Functions	G-codes	Functions
G42	Tool radius compensation (right)	G73	Peck drilling cycle
G43	Tool length compensation (positive)	G74	Counter tapping cycle
G44	Tool length compensation (negative)	G76	Fine boring cycle
G49	Tool length compensation cancel	G80	Canned cycle cancel/External operation function cancel
G50	Scaling cancel	G81	Drilling cycle
G51	Scaling	G82	Drilling cycle or counter boring cycle
G52	Local coordinate system setting	G83	Peck drilling cycle
G53	Machine coordinate system selection	G84	Tapping cycle
G54	Workpiece coordinate system 1 selection	G85	Rough boring cycle
G55	Workpiece coordinate system 2 selection	G86	Boring cycle
G56	Workpiece coordinate system 3 selection	G87	Back boring cycle
G57	Workpiece coordinate system 4 selection	G90	Absolute programming
G58	Workpiece coordinate system 5 selection	G91	Increment Cprogramming
G59	Workpiece coordinate system 6 selection	G92	Set workpiece coordinate system/ Clamp at maximum spindle speed
G65	Macro call	G94	Feed per minute
G66	Macro modal call	G95	Feed per rotation
G67	Macro modal call cancel	G96	Constant surface speed control
G68	Coordinate rotation/Three dimensional coordinate conversion	G97	Constant surface speed control cancel
G69	Coordinate rotation cancel/Three dimensional coordinate conversion cancel	G98	Return to initial point in canned cycle

1. G00：Rapid Positioning

Format：G00 X_Y_Z_

2. G01：Linear Interpolation

Format：G01 X_Y_Z_F_

3. G02 (CW) /G03 (CCW)：Circular Arc Interpolation or Helical Interpolation

Format：G02 X_Y_R_F_(G17:X_Y_,G18:X_Z_,G19:Y_Z_)

G02 X_Y_I_J_F_

The end point of an arc is specified using the addresses X，Y，or Z，and is expressed as either an absolute or an incremental value，depending on whether G90 or G91 is used. For the incremental value，the distance to the end point is specified relative to the start point of the arc.

The arc center is specified using the addresses I, J, K for X, Y, and Z axes, respectively.

4. G04: Dwell

Format: G04 X_ or G04 P_ (Dwell with the time in seconds or speed specified)
For dwell with the speed specified, another option is required.
For example, G04 X3.5 or G04 P3500 means the dwell is 3.5s.

5. G40, G41, G42: Cutter Compensation

Format: G40 G01 X_Y_
G41(G42)G01 X_Y_D_

Where, G41—Tool radius compensation left.

G42—Tool radius compensation right.

G40—Tool radius compensation cancel.

D—Code for specifying as the tool compensation value (1—3 digits).

When the tool moves, the tool path can be offset by the tool radius.

To make an offset equal to the tool radius, the CNC system first creates an offset vector that has a length matching the tool radius. The offset vector is perpendicular to the tool path. The tail of the vector is positioned on the workpiece side, while the head points towards the center of the tool.

6. G92: Setting for Workpiece Coordinate System or Clamp at Maximum Spindle Speed

Format: G92 X_Y_Z_

A workpiece coordinate system is set by specifying a value after G92 in the program.
G92 example (Figure 2.37):

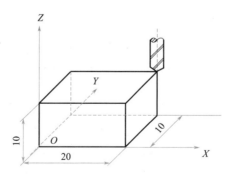

Figure 2.37 G92 Example

G92 X20. Y10. Z10. ;

7. G53: Machine Coordinate System Selection

Format: G53 G90 X_Y_Z_

When a command specifies a position using the machine coordinate system, the tool moves to the position by rapid traverse. G53, which is used to select the machine coordinate system, is a one-shot G-code; that is, it is only valid in the block in which it is specified.

G53 example (Figure 2.38):

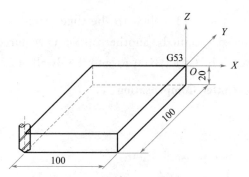

Figure 2.38　G53 Example

```
G53 G90 X-100. Y-100. Z-20.;
```

8. G54/G55/G56/G57/G58/G59: Workpiece Coordinate System 1—6 Selection

Format: `G54 G90 G00(G01)X_Y_Z_(F_)`

Workpiece coordinate system 1 to 6 are established after the reference position is returned, and the power is turned on. When the machine tool is powered on, the G54 coordinate system is automatically selected by default. You can choose from any of the six workpiece coordinate systems using the CRT/MDI panel.

Workpiece coordinate system selection example (Figure 2.39):

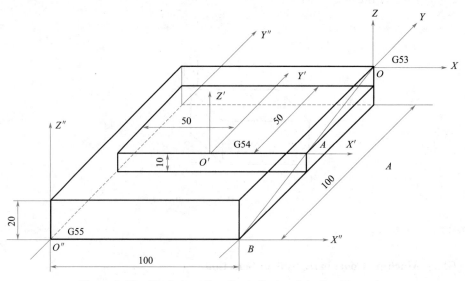

Figure 2.39　Workpiece Coordinate System Selection Example

N10 G53 G90 X0. Y0. Z0. ;
N20 G54 G90 G01 X50. Y0. Z0. F100;
N30 G55 G90 G01 X100. Y0. Z0. F100;   (Trajectory of tool nose—OAB)

G55 example (Figure 2.40):

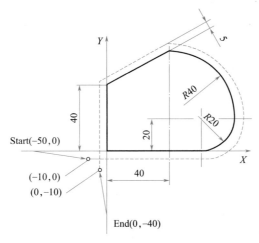

Figure 2.40 G55 Example

O24401
N10 G55 G90 G01 Z40. F2000;   (Workpiece coordinate system 2 selection)
N20 M03 S900;
N30 G01 X-50. Y0. ;   (Start of X,Y)
N40 G01 Z-5. F100;   (Start of Z)
N50 G01 G42 X-10. Y0. H01;   (Tool radius compensation right)
N60 G01 X60. Y0. ;   (Cut in)
N70 G03 X80. Y20. R20. ;
N80 G03 X40. Y60. R40. ;
N90 G01 X0. Y40. ;
N100 G01 X0. Y-10.   (Cut out)
N110 G01 G40 X0. Y-40. ;   (Tool compensation cancel)
N120 G01 Z40. F2000;
N130 M05;
N140 M30;   (End)

9. G68/G69: Coordinate Rotation

Format: G68 X_Y_R_

Where, X, Y—Absolute command for two of the X, Y, and Z axes that correspond to the current plane selected by a command (G17, G18, or G19). The command specifies the coordinates of the rotation center for the values.

R—Angular displacement with a positive value indicates counterclockwise rotation.

G69 is used to cancel a coordinate rotation.

A programmed shape can be rotated using a specific function. This allows for easy modification of a program when a workpiece is placed at an angle different from the original programmed position on the machine tool. Furthermore, if there is a pattern comprising identical shapes at various rotated positions, you can reduce both the programming time and the length of the program by creating a subprogram of the shape and calling it after applying the necessary rotations.

Coordinate rotation example (Figure 2.41):

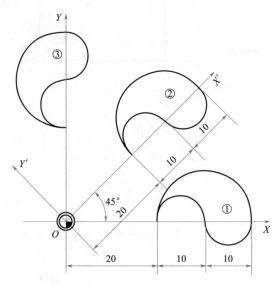

Figure 2.41　Coordinate Rotation Example

```
O24402                    (Main program)
N10 G90 G17 S900 M03;
N20 M98 P1100;            (①)
N30 G68 X0. Y0. P45;      (Coordinate rotation 45°)
N40 M98 P1100;            (②)
N50 G69;                  (Coordinate rotation cancel)
N60 G68 X0. Y0. P90;      (Coordinate rotation 90°)
M70 M98 P1100;            (③)
N80 G69 M05 M30;          (Coordinate rotation cancel)


Subprogram(①)
O1100
N100 G90 G01 X20. Y0. F100;
N110 G02 X30. Y0. R5. ;
N120 G03 X40. Y0. R5. ;
N130 X20. Y0. R10. ;
```

N140 G00 X0. Y0. ;
N150 M99;

10. G90/G91: Absolute Programming/Incremental Programming

There are two ways to command the tool's movements: absolute programming and incremental programming. In the absolute programming, the coordinate value of the end position is programmed. In the incremental programming, the movement distance from the current position is programmed.

11. M98/M99: Subprogram Call

Format: M98 P_

Where, P—Number of times the subprogram is called repeatedly.

When no repetition data is specified, the subprogram is called just once.

【拓展视频】

M99 is used to end a subprogram.

12. G50/G51: Scaling

Format: G51 X_ Y_ Z_ P_

Where, X, Y, Z—Absolute programming for center coordinate value of scaling.

P—Scaling magnification (0.001—999.999).

13. G24/G25: Programmable Mirror Image

Mirror image example (Figure 2.42):

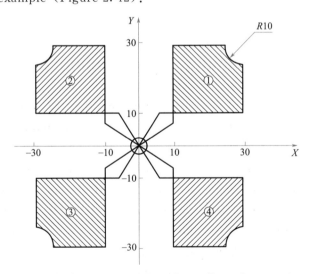

Figure 2.42　Mirror Image Example

O24403　　　　　　　　　(Main program)
N10 G91 G17 S900 M03;
N20 M98 P1101;　　　　　(①)
N30 G24 X0. ;　　　　　　(Y axis, X=0)

```
N40 M98 P1101;              (②)
N50 G24 X0. Y0. ;           (X axis,Y axis,(0,0))
N60 M98 P1101;              (③)
N70 G25 X0. ;               (Cancel a programmable mirror image Y)
N80 G24 Y0. ;               (X axis,Y0)
N90 M98 P1101;              (④)
N100 G25 Y0. ;              (Cancel a programmable mirror image)
N110 M05;
N120 M30;

Subprogram(①):
%1101
N200 G41 G00 X10. 0 Y4. 0 D01;
N210 Y1. 0
N220 Z-98. 0;
N230 G01 Z-7. 0 F100;
N240 Y25. 0;
N250 X10. 0;
N260 G03 X10. 0 Y-10. 0 I10. 0;
N270 G01 Y-10. 0;
N280 X-25. 0;
N290 G00 Z105. 0;
N300 G40 X-5. 0 Y-10. 0;
N310 M99;
```

2.5 Computer Aided Manufacturing

In the manual preparation of CNC programming, the CNC programmer must define the machine tool or tool movements in numerical terms. For complex 3D surfaces, manual programming is often not feasible.

Over the past years, sufficient efforts have been devoted to automate the generation of programming. With the development of CAD/CAM systems, interactive graphic systems have been integrated with CNC programming. Graphic-based software, using menu-driven techniques, improves user-friendliness. The CNC programmer can either create the geometrical model within the CAM package or directly extract it from the CAD/CAM database. Built-in tool motion commands assist the CNC programmer in automatically calculating tool paths. These paths can be verified through graphic display using the CAM system's animation function, greatly enhancing both the speed and accuracy of tool path generation.

The work flow needed to produce a part using an CNC machine tool can be summarized as Figure 2.43. The tasks can be classified into three main categories:

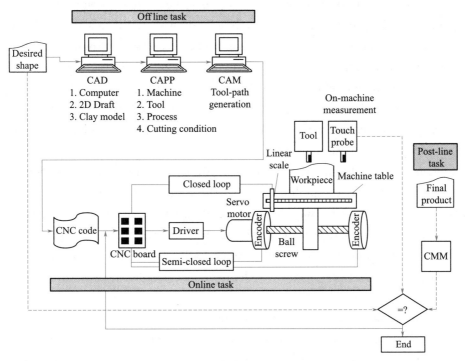

Figure 2.43 The Architecture of CNC Machine Tools and Machining Operation Flow

(1) Offline tasks: CAD, CAPP, CAM.

Offline tasks are essential for generating a program to control a CNC machine tool. In the offline stage, once the shape of a part is decided, a geometry model of this part is created using 2D/3D CAD. CAD in this context refers to a modelling stage in which both design and analysis are included because engineering analysis cannot be carried out on the shop floor.

After finishing geometric modelling, CAPP (computer aided process planning) is carried out. This stage involves generating the necessary information for machining, such as selecting machine tools, tools, jigs, and fixtures, as well as decisions about cutting conditions, scheduling, and machining sequences. Because of the complex process planning and the immature CAPP with respect to technologies, process planning generally depends on the know-how of a process planner.

CAM is executed in the final stage for generating a program. Tool paths are created based on the geometry information from CAD and the machining information from CAPP. During tool path generation, considerations interferences, minimizing machining time and tool changes, and optimizing machine performance are taken into account. In particular, CAM is an essential tool to generate 2.5D or 3D tool paths for machine tools with more than three axes.

(2) Online tasks: CNC machining, monitoring, and On-machine measurement.

Online tasks are essential machining parts using CNC machine tools. A program, which consists of machine-readable instructions, is typically generated in the offline stage. For simple parts, the program can be directly edited on the CNC machine tools by the user. In the online stage, the CNC system reads and interprets the programs from its memory, controlling the movement of the machine tool's axes. The CNC system generates instructions for position and velocity control based on the programs, and servo motors are controlled based on the instructions generated. As the rotation of servo motor is converted into linear movement via ball-screw mechanisms, the workpiece or the tool is moved, and the part is machined by these movements.

To increase the machining accuracy, not only the accuracy of the servo motor, table guide, ball screw, and spindle, but also the rigidity of the machine's construction should be high. The machine and its components must be designed to minimize insensitive to vibration and temperature fluctuations. In addition, the performance of the encoders and sensors included in the CNC system, and the control mechanism, influence the machining accuracy. The control mechanism will be addressed in more detail in the following section.

In the online stage, the state of the machine tool and the machining process can be monitored in real time. Techniques, such as tool-breakage detection, compensation of thermal deformation, adaptive control, and compensation for tool deflection based on monitoring cutting force, heat, and electric current, are applied during machining. On-machine measurement is also used to calculate machining errors by inspecting the finished part directly on the machine tool, these machining errors can then be fed back into the CNC system for compensation.

(3) Post-line tasks: Computer-aided inspection (CAI) and post-operation tasks.

The post-line task involves performing CAI on the finished part. In this stage, inspection using a CMM (coordinate measurement machine) is used to make a comparison between the machined part and the original geometry model. Any discrepancies by modifying cutter compensation or by doing post-operations such as remachining or grinding. This stage also includes reverse engineering, where the shape of the part is measured, and a geometric model based on the measured data is generated. As mentioned above, these three stages enable CNC machine tools to achieve high accuracy and productivity. CNC machine tools are capable of machining parts with complex shapes as well as simple shapes. Because they can easily switch between different programs and repetitively machine the same-shaped part shape by storing the program, CNC machine tools are highly versatile and suitable for general-purpose use.

There are several CAM systems available in the market. The flow chart of a CAM system is shown in Figure 2.44. Their basic features can be summarized as follows:

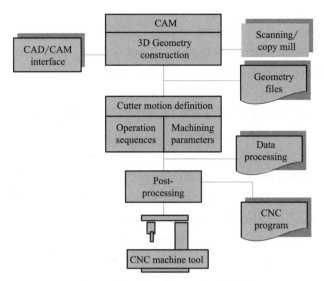

Figure 2.44 The Flow Chart of a CAM System

(1) Geometric modeling/CAD interface.

The geometry of the workpiece can be defined by basic geometrical elements such as points, lines, arcs, splines, or surfaces. These two-dimensional or three-dimensional geometrical elements are stored in the computer's memory as mathematical models. The models can be in the form of a wireframe model, a surface model, or a solid model.

In addition, geometric models can be imported from other CAD/CAM systems through standard CAD/CAM interface formats, such as the Initial Graphic Exchange Specification (IGES). IGES is a graphic exchange standard jointly developed by industry and the National Bureau of Standards with the support of the U.S. Air Force. It facilitates the transfer of 3D geometry data between different systems. This system allows for the translation of geometrical elements from one system into a neutral file standard, which can then be converted into other formats.

(2) Tool motion definition.

After geometric modelling, machining data, such as job setup, operation setup, and motion definition, is input into the computer to produce the cutting location file (CL file) for machining the workpiece.

① Job setup: It is to input machine datum, home position, and tool diameters into the CL file.

② Operation setup: It is to input operation parameters such as feed rate, tolerance, approach/retract planes, spindle speed, coolant on/off, stock offset, and tool selection into the system.

③ Motion definition: Built-in machining commands are used to control the tool's motion to machine the products. This includes processes such as profiling, pocketing, surface machining, and gouge checking.

(3) Data processing.

The input data is translated into a format usable by the computer. The computer processes the desired part surface, calculates the tool offset surface, and finally computes the paths of the tool, which are stared as the CL file. The tool paths can normally be animated graphically on the display for verification purposes.

Additionally, production planning data, such as tool list, setup sheets, and machining times, are calculated for the user's reference.

(4) Post-processing.

Different CNC machine tools have varying features and capabilities, leading to differences in the format of CNC programs. A process known as post-processing is required to convert the general instructions from the tool location file into a specific format suitable for a particular machine tool. A post-processor is a software tool that converts the tool location data files into a format that the machine controller can interpret correctly. Generally, there are two types of post-processors:

① Specific post-processor: This is a tailor-made software that outputs precise code for a specific CNC machine tool. The user is not required to change anything in the program.

② Generic (universal) post-processor: This is a set of generalized rules that the user must customize to meet the requirements of a specific CNC machine tool.

(5) Data transmission.

After post-processing, the CNC program can be transmitted to CNC machine tools either through offline or online processes.

① Offline processes: Data carriers such as paper tapes, magnetic tape, or magnetic discs, are used to transmit the CNC program to CNC machine tools.

② Online processes: Online processes is commonly used in DNC operations, where data are transferred either serially or parallelly using data cables.

(6) Serial transmission.

A synchronous serial transmission is the most widely used method, with RS-232C being the most popular synchronous standard. Many computers are equipped with a built-in RS-232C serial port (9 pins or 25 pins). RS-232C is inexpensive, easy to program, and supports baud rates up to 38400, though its noise margin is limited to distances of up to 15m.

Parallel transmission is commonly used for data transmission between computers and external devices such as sensors, PLCs, or actuators. One common standard for this is IEEE 488, which includes a 24-line bus (8 for data, 8 for controls, and 8 for ground). It can transfer data at speeds up to 1 Mbps over a 20m cable.

To ensure that the CAD/CAM facilities operate smoothly, it is desirable for these facilities to be interconnected. In a local area network (LAN), terminals can access any computers on the network or devices on the shop floor without needing a physical hardwire connection, achieving speeds of up to 300 Mbps/s. For instance, Ethernet technology operates at 100Mbps/s, significantly faster than RS-232 serial communication, which typically supports speeds of 115. 2kbps/s.

The LAN consists of both software and hardware components, governed by a set of rules known as the protocol. The software design facilitates data management and error recovery, while the hardware generates and receives signals, and the media carries these signals. The protocol defines the logical, electrical, and physical specifications of the network. For an effective communication within the network, all devices must adhere to the same protocol.

Exercises

1. Read and match each of the following pictures to the name of hand tools (Figure 2.45).

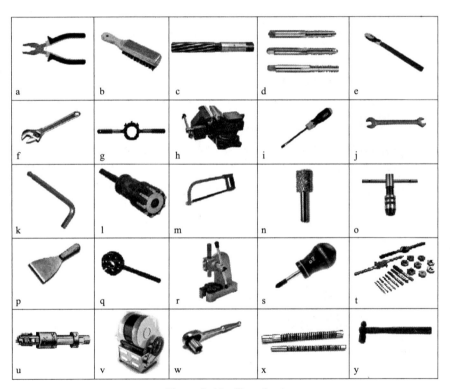

【拓展图文】

Figure 2.45   Exercise 1

[ ] hacksaw   [ ] bench vise   [ ] file card   [ ] scraper   [ ] tap and die
[ ] hand tap   [ ] hammer   [ ] burnishing tool   [ ] adjustable wrench   [ ] tap extractor
[ ] hex key   [ ] arbor press   [ ] T-handle tap wrench   [ ] hand reamer   [ ] screwdriver
[ ] plier   [ ] external lap   [ ] rotary files   [ ] Phillips screwdriver   [ ] open-end wrench
[ ] broach   [ ] socket wrench   [ ] tumbling machine   [ ] die stock   [ ] file

2. Learn the structure of vernier calipers (Figure 2.46) and give the reading of Figure

2.47.

Vernier calipers are precision tools used to make accurate measurements within 0.001 inch or 0.02 mm. They can measure the outside diameter or width of an object, the inside diameter or width of an object, and the depth of an object. Vernier calipers are available in both inch and metric graduations, and some models include both types of graduations on the same caliper.

The vernier caliper consists of an L-shaped frame and a movable jaw. The L-shaped frame includes a bar with a fixed jaw attached. The bar shows the main scale graduations. The movable jaw slides along the bar and contains the vernier scale. An adjusting nut allows for fine turning of the movable jaw, and locking screws secure the reading in place to prevent any changes during measurement.

Most calipers have bars gradutated on both sides or edges. One side is used for outside measurements, while the other side is used for inside measurements. The internal jaws are designed to measure inside spaces, such as the diameter of a hole, while the external jaws measure outside dimensions, such as the diameter of a piece of pipe.

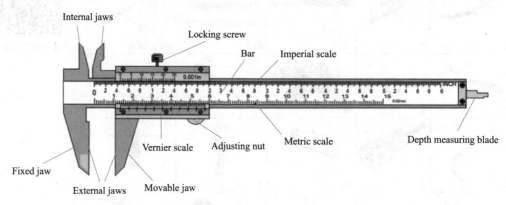

Figure 2.46　Exercise 2 (1)

Figure 2.47　Exercise 2 (2)

The reading is _____

3. Look at the following micrometer and fill in the blanks (Figure 2.48 and Figure 2.49).

The micrometer is a precision measuring instrument used by machinists. Each revolution of the ratchet moves the spindle face 0.5 mm towards the anvil face. The object to be measured is placed between the anvil face and the spindle face. The ratchet is turned clockwise until the object is 'trapped' between these two surfaces, at which point the ratchet makes a 'clicking' noise. This indicates that the ratchet cannot be tightened any further, and the measurement can be read.

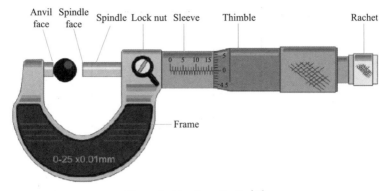

Figure 2.48 Exercise 3 (1)

Figure 2.49 Exercise 3 (2)

Sleeve reads full mm = _____.
Sleeve reads 1/2 mm = _____.
Thimble reads = _____.
Total measurement = _____.

4. Narrate CNC coordinate systems.

5. Give the CNC program of Figure 2.50 (blank dimension: $\phi 20 \times 100$, units: mm).

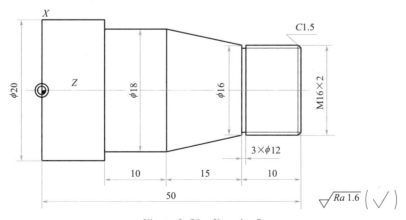

Figure 2.50 Exercise 5

6. Give the CNC program of Figure 2.51 (blank dimension: φ60×105, units: mm).

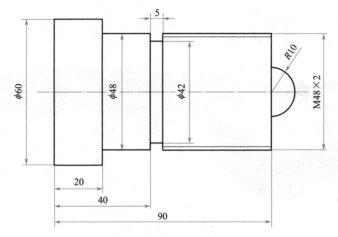

Figure 2.51　Exercise 6

7. Give the CNC program of Figure 2.52 (blank dimension: 150×150×20, units: mm).

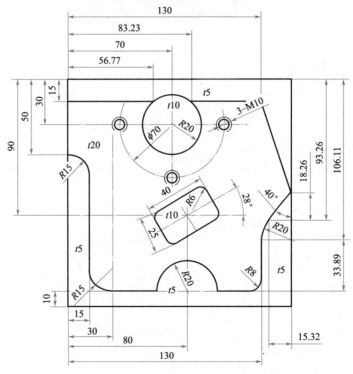

Figure 2.52　Exercise 7

8. Give the CNC program of Figure 2.53 (blank dimension: 240×160×20, units: mm).

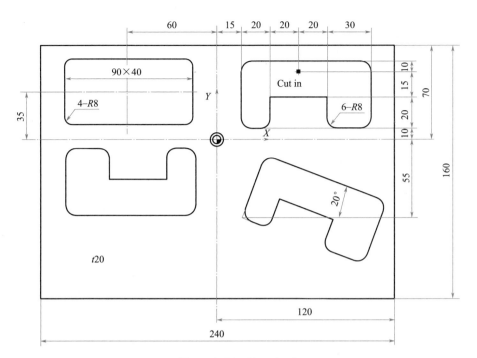

Figure 2.53　Exercise 8

Chapter 3
CNC Unit and Control Principle

Objectives

- To understand the hardware structure of the CNC unit.
- To understand CNC system software.
- To understand the interpolation principle.
- To master the cutter compensation principle.
- To understand the CNC acceleration/deceleration control.

3.1 Hardware Architecture of a CNC Unit

CNC unit is the core of CNC system. The hardware comprising a CNC unit is made of microprocessors, electronic memory modules, I/O interfaces, and position control modules, etc., which is just like an ordinary computer system. A CNC unit controls the machine tool through MCU, position transducers, tool holding device, work holding device and machine tool body (Figure 3.1).

Figure 3.2 shows data flow between modules in the CNC system. The data is transmitted between modules via ring buffers defined by global variables, and the ring buffers are located between the interpreter and the rough interpolator, between the rough interpolator and the acceleration/deceleration (abbreviation: acc/dec) controller, and between the acc/dec controller and the fine interpolator. Each ring buffer includes data. The fine interpolator and the position controller use global variables to send necessary data.

Features of the CNC system:

(1) Storage of more than one program: With improvements in computer storage technologies, newer CNC controllers have sufficient capacity to store multiple programs. CNC controller manufacturers generally offer one or more memory expansions for MCU as options.

CNC Unit and Control Principle Chapter 3

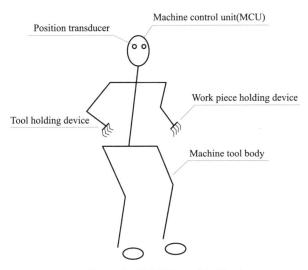

Figure 3.1 A CNC Unit and its Devices

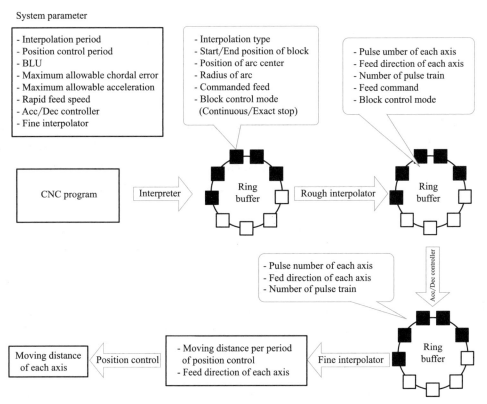

Figure 3.2 Data Flow between Modules in the CNC System

(2) Various forms of program input: Whereas conventional (hard-wired) MCUs are limited to punched tape as the input medium for entering programs, CNC controllers generally possess multiple data entry capabilities, such as punched tape, magnetic tape, floppy diskettes, RS-232 communications with external computers, and manual data input

(operator entry of programs).

(3) Program editing at the machine site: The CNC system permits a program to be edited while it resides in the MCU computer memory. Hence, a program can be tested and corrected entirely at the machine site, rather than being returned to the programming office for corrections. In addition to program corrections, editing permits cutting conditions in the machining cycle to be optimized. After the program has been corrected and optimized, the revised version can be stored on the punched tape or other media for future use.

(4) Fixed cycles and programming subroutines: The increased memory capacity and the ability to program the control computer provide opportunities to store frequently used machining cycles, such as macros, that can be called by the part program. Instead of writing full instructions for a particular cycle into every program, a programmer includes a call statement in the program to indicate that the macro cycle should be executed. These cycles often require certain parameters to be defined. For example, in a bolt hole circle, the diameter of the bolt circle, the spacing of the bolt holes, and other parameters must be specified.

(5) Interpolation: Some interpolation schemes are normally executed only on a CNC system because of computational requirements. Linear and circular interpolations are sometimes hard-wired into the control unit, but helical, parabolic, and cubic interpolations are usually executed by a stored program algorithm.

(6) Positioning features for setup: Setting up the machine tool for a given work part involves installing and aligning a fixture on the machine tool table. This must be accomplished so that the machine axes are established with respect to the work part. The alignment task can be facilitated using certain features made possible by software options in the CNC system. Position set is one of these features. With position set, the operator is not required to locate the fixture on the machine tool table with extreme accuracy. Instead, the machine tool axes are referenced to the location of the fixture using a target point or set of target points on the work part or fixture.

(7) Tool length and size compensation: In older style controls, cutter dimensions had to be set precisely to agree with the tool path defined in the program. Alternative methods for ensuring accurate tool path definition have been incorporated into the CNC controls. One method involves manually entering the actual tool dimensions into the MCU. These actual tool dimensions may differ from those originally programmed. Compensations are then automatically made in the computed tool path. Another method involves the use of a tool length sensor built into the machine tool. In this technique, the tool is mounted in the spindle, and the sensor measures its length. This measured value is then used to correct the programmed tool path.

(8) Acceleration and deceleration calculations: This feature is applicable when the tool moves at high feed rates. It is designed to avoid tool marks on the work surface that

would be generated due to machine tool dynamics when the tool path changes abruptly. Instead, the feed rate is smoothly decelerated in anticipation of a tool path change and then accelerated back up to the programmed feed rate after the direction change.

(9) Communications interface: With the trend toward interfacing and networking in plants today, most modern CNC controllers are equipped with a standard RS-232 or other communications interface to link the machine tool to other computers and computer-driven devices. This is useful for various applications, such as:

① Downloading programs from a central data file.

② Collecting operational data, such as workpiece counts, cycle times, and machine utilization.

③ Interfacing with peripheral equipment, e. g., robots that unload and load parts.

(10) Diagnostics: Many modern CNC systems possess diagnostics capabilities, such as monitoring certain aspects of the machine tool to detect malfunctions, impending malfunctions, and diagnosing system breakdowns.

3.1.1 MCU for CNC

The MCU is the hardware that distinguishes CNC from conventional NC. The general configuration of the MCU in a CNC system is illustrated in Figure 3.3.

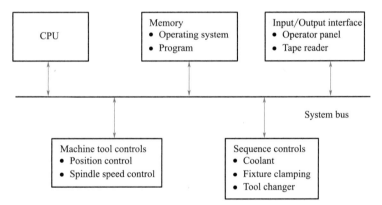

Figure 3.3 The General Configuration of the MCU in a CNC System

The MCU consists of the following subsystems:
(1) CPU.
(2) Memory.
(3) Input/Output interface.
(4) Machine tool controls.
(5) Sequence controls for other machine tool functions.

These subsystems are interconnected by means of a system bus, which communicates data and signals among the components of the network. The MCU is the heart of a CNC system. There are two subunits in the MCU: the data processing unit (DPU) and the control loop unit (CLU).

1. CPU

The CPU is the brain of the MCU. It manages other components in the MCU based on software contained in the main memory. The CPU can be divided into three sections: control section, arithmetic logic unit (ACU), and immediate access memory.

The control section retrieves commands and data from memory and generates signals to activate other components in the MCU. In short, it sequences, coordinates, and regulates all the activities of the MCU. The ALU consists of the circuitry to perform various calculations (addition, subtraction, and multiplication), counting, and logical functions required by software residing in memory. The immediate access memory provides temporary storage of data, which is processed by the CPU, and it is connected to the main memory of the system bus.

2. Memory

The immediate access memory in the CPU is not intended for storing CNC software. A much greater storage capacity is required for various programs and data to operate the CNC system. CNC memory can be divided into two categories: main memory and secondary memory. Main memory (also known as primary storage) consists of ROM and RAM devices. Operating system software and machine interface programs are generally stored in ROM. These programs are usually installed by the manufacturer of the MCU. Programs are stored in RAM devices. Current programs in RAM can be erased and replaced by new programs as jobs change.

High-capacity secondary memory (also called auxiliary storage or secondary storage) are used to store large programs and data files, which are transferred to main memory when needed. Secondary memory consists of hard disks and portable devices that have replaced most of the punched tape traditionally used to store programs. Hard disks are high-capacity storage devices that are permanently installed in the control unit of the CNC machine tool.

3. Input/Output (I/O) Interface

The I/O interface provides communication software between the various components of CNC system, other computer systems, and the machine tool operator. The I/O interface transmits and receives data and signals from external devices. The operator control panel is the basic interface, through which the machine tool operator communicates with the CNC system. This is used to enter commands related to program editing, MCU operating mode (e.g., program control vs. manual control), speeds and feeds, cutting fluid pump on/off, and similar functions. Either an alphanumeric keypad or keyboard is usually included in the operator control panel. The I/O interface also includes a display (CRT or LED) for communicating data and information from the MCU to the machine tool operator. The display is used to indicate the current state of the program as it is being executed and to warn the operator of any malfunctions in the CNC system.

As indicated previously, CNC programs are stored in various ways. Programs can also be entered manually by the machine tool operator or stored at a central computer site and transmitted via LAN to the CNC system. Whichever means is employed by the plant, a suitable device must be included in the I/O interface to allow input of the program into MCU memory.

4. Controls for Machine Tool Axes and Spindle Speed

These are hardware components that control the position and velocity (feed rate) of each machine tool axis as well as the rotational speed of the machine tool spindle. The control signals generated by the MCU must be converted to a form and power level suited to the particular position control systems used to drive the machine tool axes. Positioning systems can be classified as open-loop and closed-loop positioning systems, and different hardware components are required in each case.

Depending on the type of machine tool, the spindle is used to drive either workpiece or a rotating tool. Turning exemplifies the first case, whereas milling and drilling exemplify the second. Spindle speed is a programmed parameter for most CNC machine tools. Spindle speed components in the MCU usually consist of a drive control circuit and a feedback sensor interface. The particular hardware components depend on the type of spindle drive.

5. Sequence Controls for Other Machine Tool Functions

In addition to controls of table position, feed rate, and spindle speed, several functions are accomplished under program control. These auxiliary functions are generally on/off (binary) actuations, interlocks, and discrete numerical data. To avoid overloading the CPU, a PLC is sometimes used to manage the I/O interface for these auxiliary functions.

3.1.2 New CNC Control Technologies

1. Graphics Processing Unit

A graphics processing unit (GPU), also occasionally called a visual processing unit (VPU), is a specialized electronic circuit designed to rapidly manipulate and alter memory to accelerate the creation of images in a frame buffer intended for output to a display. GPUs are used in embedded systems, mobile phones, PCs, workstations, and game consoles. Modern GPUs are very efficient in manipulating computer graphics, and their highly parallel structure makes them more effective than general-purpose CPUs for algorithms where processing large blocks of data is done in par-

【拓展视频】

allel. In PLs, GPUs can be present on video cards, motherboards, or certain CPU dies.

The term GPU was popularized by NVIDIA Corporation in 1999, which marketed the GeForce 256 as the world's first 'GPU', a single-chip processor with integrated transform, lighting, triangle setup/clipping, and rendering engines that are capable of processing a minimum of 10 million polygons per second. Rival ATI Technologies coined the term VPU with the release of the RADEON 9700 in 2002.

Modern GPUs use most of their transistors to do calculations related to 3D computer graphics. They were initially used to accelerate the memory-intensive work of texture mapping and rendering polygons, later adding units to accelerate geometric calculations, such as the rotation and translation of vertices into different coordinate systems. Recent developments in GPUs include support for programmable shaders that can manipulate vertices and textures with many of the same operations supported by CPUs, oversampling and interpolation techniques that can reduce aliasing, and very high-precision color spaces. Because most of these computations involve matrix and vector operations, engineers and scientists have increasingly studied the use of GPUs for non-graphical calculations. An example of GPUs being used non-graphically is the generation of Bitcoins, where GPUs are used to solve hash functions.

In addition to the 3D hardware, today's GPUs include basic 2D acceleration and frame buffer capabilities (usually with a VGA compatibility mode). Newer cards like RTX 5000 adopt AMPERE architecture and DLSS (deep learning super sampling) technology.

2. Motion Controllers

A motion controller controls the motion of some objects. Frequently, motion controllers are implemented using digital computers, and they can also be implemented with only analog components.

The role of a motion controller is to control the high-speed and high-precision motion of a linear motor in accordance with instructions from a CNC unit. It holds the industrial property rights to copy and use the proportion integration differential (PID) control that realizes high-speed and high-acceleration control of a linear motor. The technologies for manufacturing a motion controller of a linear motor use modern control theories and software to control a linear motor.

In addition to the development of CNC units which dictate the overall performance of a machine tool, it is of great importance for enhancing machining precision by giving and transmitting instructions as accurately as possible to the linear motors. What is needed to calculate the motion of a linear motor accurately and transmit the calculation results to the CNC units is a motion controller. The development of the motion controller started with the aim of working out tailored units designed specifically for machine tools instead of modifying existing ones.

3. Multi-axis Motion Control Technology

In a multi-axes motion controller (Figure 3.4), the motions of individual axis must often needed to be coordinated. A simple example would be a robot that needs to move two joints to reach a new position. We could extend the motion of the slower joints so that the motion of each joint begins and ends together.

Motion controllers require a load (to be moved), a prime mover (to cause the load to move), sensors (to sense the motion and monitor the prime mover), and a controller (to provide the intelligence to cause the prime mover to move the load as desired). Motion controllers are used to achieve desired benefits which can include:

(1) Increased position and speed accuracy.

Figure 3.4　Multi-axes Motion Controller

(2) Higher speed.

(3) Faster reaction time.

(4) Increased production.

(5) Smoother movement.

(6) Reduction in cost.

(7) Integration with other automation.

(8) Integration with other processes.

(9) Ability to convert desired specifications into motion required to produce a product.

(10) Increased information and ability to diagnose and troubleshoot.

(11) Increased consistency.

(12) Improved efficiency.

(13) Elimination of hazards to humans or animals.

3.2　CNC System Software

A CNC machine tool consists of two major components: the machine tool and the controller (or the CNC unit) which is an onboard computer that may or may not be manufactured by the same company. General Numeric, Fanuc, General Electric, DMG, Cincinnati Milacron, and Siemens are among the manufacturers of CNC controllers that supply units to machine tool makers. Each controller is manufactured with a standard set of building codes, i.e., startup program, basic system codes, machining and measuring loop programs, etc. Other codes are added by the machine tool builders. Therefore, program codes vary somewhat between machine tools. Every CNC machine tool, regardless of manufacturers, is a collection of systems coordinated by the controller.

The CNC system software (Figure 3.5) can be categorized into two groups: user software and system software.

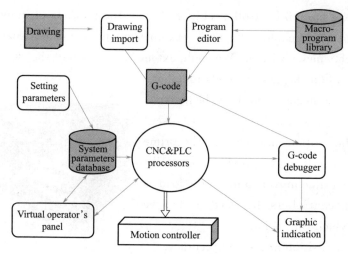

Figure 3.5　The CNC System Software

1. User Software

User software, also called the program, is programmed and input using numerical control language (such as APT) to indicate the machining procedure of the part. It compiles into a program with various G-code and M-code arranged in process sequence.

2. System Software

The primary functions of a CNC system are mainly brought out by the system software. System software performs the following tasks through corresponding subroutines:

(1) Compiling the part codes input by users.

After receiving the part codes in ISO or EIA format, the compiler translates, trims, and stores them into a specified format, decodes the machining instructions, performs decimal-binary transitions to coordinate data, calculates tool's center path considering the tool radius offset, and pre-calculates some constants for use in interpolation calculation and speed control processes. While preprocessing steps vary between CNC systems, they all aim to save time during real-time interpolation calculations. Preprocessing input part codes in real-time is not crucial and can be done before machining or during idle intervals while machining. The more thorough the preprocessing of inputted part codes, the smoother the real-time calculations of interpolation and speed control will be.

(2) Interpolation calculations.

The interpolation calculation subroutine in a CNC system functions is similar to a hardware interpolator, as it generates electrical pulses for the axes. It is a crucial real-time program that requires minimal instruction codes to reduce the time needed for interpolation calculations, which directly influences the feed speed. In some CNC systems, a combination of rough and fine interpolation is used: software handles rough interpolation by calculating each time, and then hardware performs fine interpolation by converting the segment into a series of pulses and outputting them. This method enhances feed speed and frees the computer from complex interpolation calculations, allowing it to focus on other necessary processes.

(3) Tool compensation.

CNC machine tools require various forms of compensation, such as tool length compensation, tool radius compensation, and tool nose radius compensation. While these compensations serve different purposes depending on the machining type, all compensations allow the CNC operator to modify for unpredictable conditions related to tooling. Generally speaking, if the CNC operator encounters any unpredictable situations during programming, the CNC control manufacturer has likely developed a form of compensation to deal with the issue.

(4) Speed control.

The speed control subroutine is responsible for managing the timing of pulse generation, specifically controlling the frequency of interpolation calculations according to the specified speed code (or corresponding speed instruction) to maintain the preset feed velocity. In the event of an unexpected abrupt change in velocity during machining, the speed control subroutine should automatically adjust speed either speeding up or slowing down, to avoid loss of pace in the drive system.

Speed control can be implemented totally by software methods (software timer methods) or by hardware means. For instance, hardware implementation may involve using velocity code to control an oscillator, allowing the CNC unit to perform intermittent interpolation calculations to maintain the desired feed velocity. Additionally, by processing speed control data in software and integrating it with speed integrator hardware, high-performance constant compound velocity control can be achieved, significantly enhancing feed velocity.

(5) Position control.

Position control operates within the position loop of the servo system and can be accomplished using either software or hardware. The position control software compares the position calculated through interpolation with the actual measured position during each sampling cycle, adjusting the motor's operation based on differences. Furthermore, position control software can adjust the gain of the position loop circuit, compensate for screw pitch errors in each axis direction, and address non-return-to-zero issues when reversing motion. These adjustments contribute to improved location precision.

3.3 Interpolation Principle

Every CNC machine tool discussed in this book features more than one axis of motion. It is often necessary for programmers to command two or more axes moving simultaneously in a controlled manner. For example, when using an end mill tool to machine straight, angular, and round surfaces, some movements may involve only one axis; however, angular and circular motions require at least two axes.

Globally, many companies produce CNC systems, with Siemens (Germany), Fanuc (Japan), and Allen Bradley (America) having a great influence on China's manufacturing automation. China also has its own CNC brands widely used for manufacturing and educational purposes, such as HNC (Huazhong numerical control system) and Blue-Sky numerical control system, and JD numerical control system.

Three commonly used types of CNC systems in China are Fanuc, Siemens, and HNC.

【拓展视频】

In the early days of NC, programing angular or circular surfaces required breaking the motion into a series of very small one-axis movements to approximate the desired shape as closely as possible. Producing such motion normally requires computer assistance. However, with the advent of motion interpolation, programming complex movements has become much simpler. Today's CNC systems make it relatively easy to command angular and circular motions.

Interpolation in CNC refers to the mathematical concept of estimating missing functional values at neighboring points. When a CNC interpolates motion, it precisely estimates the programmed path based on limited input data.

Linear interpolation involves connecting any programmed points with straight lines, regardless of how close or far apart the points are. Curves can be approximated using linear interpolation by deviding them into short, straight-line segments. However, this method has limitations, as a large number of points need to be programmed to accurately describe a curve and produce a contour shape.

1. Point-by-point Comparison Method for Linear Interpolation

When a CNC control makes a straight motion involving two axes, only the start and end points of the motion are required. The control automatically and instantaneously fills in the missing points between the start and end points. In practice, the control executes a series of very small, one-axis movements from the start to the end point, creating a "staircase effect". Each step is very small, making the result appear as a straight line. As shown in Figure 3.6, when two or more axes are programmed, the control forms a series of small, one-axis movements. The resolution of the axis is determined by the size of each step: smaller steps, higher resolution.

2. Point-by-point Comparison Method for Circular Arcs Interpolation

Circular arcs interpolation employs an algorithm known as the incremental interpolator, which determines the direction of each step at regular intervals and sends pulses to the related axis. This section describes the stair-step approximation method for circular arcs interpolation, particularly when the circular movement is commanded in a clockwise direction in the first quadrant relative to the center of the circle (Figure 3.7).

$$F_m = X_m^2 + Y_m^2 - R^2 \tag{3.1}$$

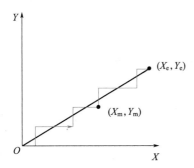 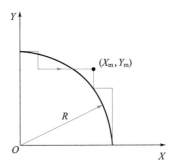

Figure 3.6 Point-by-point Comparison Method for Linear Interpolation

Figure 3.7 The Behavior of the Stair-step Approximation Interpolation Algorithm

The direction of each step is based on F_m, the commanded circular direction and the quadrant of movement. For example, if the circular movements carried out in a clockwise direction in the first quadrant, the algorithm executed is as below:

$F_m > 0$, Step $-Y$: The position (X_m, Y_m) is located on the outside of the circle. In this case, the step moves in the negative direction of Y axis.

$F_m = 0$, Step $+X$: One of the above rules can be arbitrarily selected and applied.

$F_m < 0$, Step $+X$: The position (X_m, Y_m) is located on the inside of the circle. In this case, the step moves in the positive direction of X axis.

After one step is completed by applying the above rules, the position (X_{m+1}, Y_{m+1}) is updated, and the procedure is repeated until the tool reaches the commanded position (X_f, Y_f).

3.4 Tool Compensation Principle

As we begin our discussion on the specific types of compensation available for different CNC machine tools, it is important for beginners to understand the reasons behind these compensations. Understanding the purpose of each form of compensation is crucial for effectively using it. If you understand why you need the compensation type, you will be better equipped to apply it. Also, knowing the reasons for compensation will enable you to adapt to various implementations across different machine tools and CNC systems.

Table 3-1 provides a list of types of tool compensation and CNC machine tools related to each type.

Table 3-1 The Tool Compensation Types and the Related CNC Machine Tools

Tool Radius Compensations	CNC Machine Tools
Tool length compensation	Machining centers
Tool radius compensation	Machining centers
Fixture offsets	Machining centers
Dimensional tool offsets	Turning centers
Tool nose radius compensation	Turning centers
Wire radius compensation	Wire EDM machine tools
Wire taper compensation	Wire EDM machine tools

3.4.1 Tool Compensation

Using tool compensation values simplifies the programming of a workpiece by eliminating the need to manually adjust for specific tool lengths or radii. Instead, the dimensions provided in the workpiece drawing can be directly used for programming. The CNC system automatically accounts for both tool lengths and radii.

1. Tool Length Compensation for Milling and Turning

Tool length compensation is essential for adjusting the discrepancy between the set (programmed) tool length and the actual tool length.

In the case of a milling tool, the tool length L is defined in Z direction. As shown in Figure 3.8, B—tool setup point; L—Length, the distance of the cutting tip to the tool setup point in the Z direction; R—Radius of the milling tool.

In the case of a turning tool, the tool length L is defined in the Z direction. As shown in Figure 3.9, B—Tool setup point; L—Length, the distance of the cutting tip to the tool setup point in the Z direction; Q—Overhang, the distance of the cutting tip to the tool setup point in the X direction; R—Cutting radius.

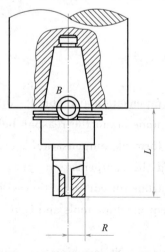

Figure 3.8 Tool Compensation Values on a Milling Cutter

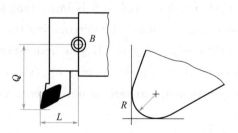

Figure 3.9 Tool Compensation Values on a Turning Tool

In CNC systems, tool compensation values are stored in a dedicated compensation value storage. Most CNC systems can accommodate up to 99 tools. These values must be activated during machining by calling the data within the CNC programs, typically using the address "H" or by specifying particular locations in the "T" word.

2. Tool Radius Compensation

Tool radius compensation (also referred to as tool diameter compensation) is used in machining centers and similar CNC machine tools. This feature allows programmers to disregard the tool's radius or diameter during programming. By using tool radius compensation, programming becomes more straightforward, as programmers do not have to concern about the exact tool diameter while the program is being prepared. Tool radius compensation also allows for variations in the tool's radius without necessitating any modifications to the program (Figure 3.10).

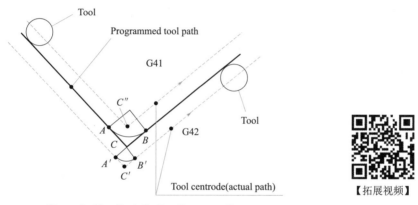

Figure 3.10 Tool Radius Compensation

(1) The work principle of tool radius compensation.

Understanding how the CNC control interprets tool radius compensation commands is crucial for solving any related issues. While there may be minor differences in how each control manufacturer handles tool radius compensation internally, the work principle discussed here apply to most modern CNC systems.

Using tool radius compensation involves three basic programming steps:

① Instate tool radius compensation.

② Make movements to machine workpiece.

③ Cancel tool radius compensation.

Tool radius compensation is instated with a command that instructs the CNC control on how to position the tool relative to the surfaces being machined during its movements. The tool can be positioned to either the left (using G41) or the right (using G42) of the surface. A simple way to remember G41 and G42 is by understanding the differences between climb milling and conventional milling. When using a right-hand tool (with the spindle rotating clockwise, typically activated with M03), G41 is used for climb milling, while G42 is used for conventional milling. Once tool radius compensation is instated, the control keeps the tool to the left or right side of the tool path defined by straight line (G01) and circular (G02 and G03) commands. These tool paths represent the actual surfaces being machined.

Tool radius compensation remains active until it is explicitly cancelled. The command to cancel tool radius compensation is G40, after which the control will no longer maintain the tool on either side of the motion commands.

(2) Limitations of tool radius compensation.

Tool radius compensation is not applicable to all types of tools. It is specifically needed for tools that have the ability to machine along their peripheries, such as end mills, shell mills, and certain face mills. In contrast, tools like drills, reamers, taps, boring bars, and other center-cutting tools do not use tool radius compensation.

There are many instances during milling operations when the programmer aims to machine the edge of a workpiece, which can take the form of either a straight surface or a contoured surface. Similar to how a woodworking router operates, a milling cutter is driven along the edge of the workpiece.

One method for programming the milling cutter's path involves using the centerline of the cutter. In this case, the programmer must account for the diameter of the milling cutter. For example, if the milling cutter has a diameter of 1 inch (2.54cm), all programmed motions must be kept precisely 0.5 inches (1.27cm) away from the surfaces to be machined. This calculation assumes that there is no cutter pressure pushing the milling cutter away from its programmed path.

While programming using the centerline coordinates of the tool path is a common practice, it comes with several limitations.

Each of these limitations posed significant challenges before the introduction of tool radius compensation. When programming based on the tool's centerline coordinates, extra calculations were required for each axis in the coordinate system to account for the tool's radius. This adds complexity and increases the likelihood of errors during programming.

3.4.2 Reasons for Tool Radius Compensation

Tool radius compensation offers several advantages that significantly ease the programming process. Here, we list and explain why utilizing this feature is so beneficial:

(1) Simplified programming.

(2) Relationship between roughing and finishing.

(3) Error compensation for the tool.

Consider a scenario where a programmer has designed a programming of a 1-inch diameter end mill to machine the right side of a rectangular workpiece. If the program is based on the tool's centerline coordinates and does not use tool radius compensation, the end mill must be kept precisely 0.5 inches away from the workpiece's edge throughout its motion.

Now, imagine that during setup, the operator discovers that 1-inch end mills are unavailable. The only available options are 0.875-inch or 1.25-inch diameter end mills. Without tool radius compensation, the programmer would have to manually modify the program to accommodate the different diameters.

Most machinists will agree that achieving the desired results with a tool on the first attempt is rare. During machining, the tool, workpiece, and even the machine tool itself are subjected to significant pressure. The more powerful the machining operation, the greater the pressure exerted. Even when using a rigid setup, such as a sturdy end mill holder and a short overall length, the tool deflection during machining is almost inevitable. This deflection occurs because the cutting edge of the tool tends to push away from the surface being machined.

If you have ever scraped paint from a wall, you have experienced this kind of deflection. When scraping paint, you do your best to keep the scraper against the wall, but many times the scraper deflects away from the surface. In machining, this tendency to deflect is called tool deflection.

Generally speaking, the weaker the machine tools and tools, the greater the potential for tool deflection. Minor tool deflection may not always cause problems, but it can be a significant issue when high accuracy is required.

For instance, in the previous example of milling the right side of a workpiece, even if an end mill with a precise 1-inch diameter is used, tool deflection is still possible. Depending on the tolerance required for the surface, this deflection may be substantial enough to cause the workpiece to be tall out of tolerance. If this occurs and the program relies on fixed centerline coordinates, it would necessitate reprogramming the milling cutter to compensate for the tool deflection.

Additionally, it is important to note that tool deflection tends to increase as the tool becomes dull. A sharp tool will experience less tool deflection than a dull one, and this change in tool deflection over the life of the tool can create significant challenges. When using fixed centerline coordinates in a program, these changes in tool deflection can lead to increased difficulties in maintaining the desired accuracy.

3.4.3 Complex Contours

Despite the potential issues discussed earlier, there are scerarios where using centerline coordinates can be justified particularly for simple workpieces. With the previous example of milling the right side of a workpiece, it would be relatively easy to simply add or subtract the tool's radius from the square sur-

【拓展视频】

face of the workpiece. However, as the surface to be milled becomes more complex, calculating the centerline coordinates for the tool of the end mill becomes increasingly challenging, when dealing with angular surfaces or radii, it can be difficult enough to calculate coordinates on the workpiece, let alone the additional complexity of calculating the centerline coordinates for the tool.

When a programmer calculates the tool's centerline coordinates for the tool path, the task can be very difficult (Figure 3.11). In this example, it would be hard enough to calculate the actual surfaces of the workpiece and the tool's centerline coordinates.

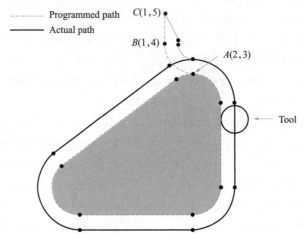

Figure 3.11　Tool Radius Compensation Entry Moves for Material Edge Contour

The reason for using tool radius compensation has something to do with roughing operations. We have already stated that, when a complex contour must be machined, it can be difficult enough to come up with workpiece coordinates, let alone tools centerline coordinates. Moreover, there exists the complication of maintaining a constant amount of finishing stock throughout the surface being machined during the roughing process. In essence, this doubles the amount of workload the programmer must do. They must not only calculate the centerline coordinates of the end mill during the finishing phase, but also calculate them for the roughing operation.

3.5　CNC Acceleration and Deceleration Control

It is important to ensure smooth movement and minimize the impact of machine tools in high-precision CNC machining. However, in practical machining, the impact and vibration on machine tools can negatively affect the cutting process and compromise machining quality. It has been reported that the unsmooth acceleration and deceleration control during feed movement is a primary cause of this impact and vibration.

(1) Interpreter function: An interpreter plays a role in reading the program, inter-

preting the ASCII blocks within it, and storing the interpreted data in internal memory for the interpolator. In general, the CNC system issues commands based on the interpreted data while the interpreter continues to read and interpret the next block. If the time taken to interpret the block exceeds the time to required finish the command, the machine tool must wait for the interpretation of the next block to finish, leading to unavoidable machine stops. Therefore, to prevent the machine tools from stopping, a buffer that temporarily stores the interpreted data is used. The buffer refers to as the internal data buffer, always keeping a sufficient number of interpreted data.

(2) Interpolator function: An interpolator plays a role in sequentially reading the data from the internal data buffer, calculating the position and velocity for each axis over time, and storing the results in a FIFO (first in, first out) buffer for the acceleration and deceleration controller. Linear and circular interpolators are typically used in CNC systems, while parabola and spline interpolators are used for certain CNC systems. The interpolator generates pulses corresponding to the tool path data (e.g. line, circle, parabola, spline) and sends these pulses to the FIFO buffer. The number of pulses is determined by the length of the tool path, while the frequency of the pulses is based on the desired velocity. In a CNC system, the displacement per pulse determines the accuracy. For example, if an axis moves 0.002 mm per pulse, the accuracy of the CNC system is 0.002 mm. In addition, the CNC system should generate 25000 pulses for moving 50 mm, which translates to 8333 pulses per second to achieve a speed of 1m per minute.

(3) Position control and vibration mitigation: If position control is executed by using data generated by the interpolator, large mechanical vibrations and shocks can occur whenever the workpiece starts or stops moving. To mitigate these issues, filtering for acceleration and deceleration control is executed before the interpolated data is sent to the position controller. This method is known as "Acceleration/deceleration After Interpolation". An alternatives method called "Acceleration/deceleration Before Interpolation" executes acceleration and deceleration control prior to interpolation.

(4) Position controller function: The data from the acceleration and deceleration controller is sent to the position controller, which performs position control based on the transmitted data at constant time intervals. Position control typically utilizes a PID (proportional-integral-derivative) controller, which issues velocity commands to the motor driving system in order to minimize the position difference between the commanded position and the actual position detected by the encoder. However, issues related to noise cannot be fully avoided when using analog signals. The acceleration, deceleration, and jerk values are specified within the CNC system or by the CNC programmer.

The linear law of acceleration and deceleration is shown in Figure 3.12.

The exponential curve law of acceleration and deceleration is shown in Figure 3.13.

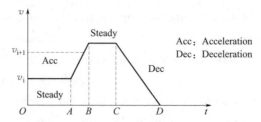

Figure 3.12　The Linear Law of Acceleration and Deceleration

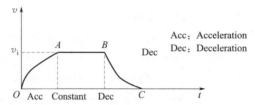

Figure 3.13　The Exponential Curve Law of Acceleration and Deceleration

3.6　PLC Function

The NCK (numerical control kernel) unit is the core of the CNC system. It interprets the program and executes interpolation, position control, and error compensation based on this interpretation. Ultimately, it controls the servo system, enabling the workpiece to be machined. The PLC sequentially controls tool change, spindle speed adjustment, workpiece change, and input/output signal processing, effectively managing the machine tool's behavior, excluding servo control.

The logic controller is used to execute sequential control in machine tools and industries. Traditionally, logic control was implemented using hardware that consisted of relays, counters, timers, and circuits, which is categorized it as a hardware-based logic controller. However, modern PLC consists of a few electrical devices, such as microprocessors and memory (Figure 3.14). It can carry out logical operations, counter functions, timer functions, and arithmetic operations. Therefore, a PLC can be defined as a software-based logic controller.

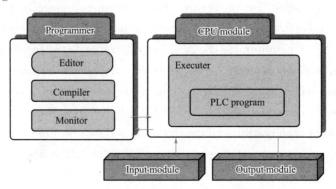

Figure 3.14　The Structure of PLC

The advantages of PLC are as follows:

(1) Flexibility: The control logic can be adjusted by modifying the program alone.

(2) Scalability: The system can expand by adding modules and changing programs.

(3) Economic efficiency: Reduction of cost is possible due to the decrease of design time. The system is highly reliable and easily maintained.

(4) Miniaturization: The physical size is smaller compared to traditional relay control boxes.

(5) Reliability: The probability of failures due to poor contacts decreases with the use of semiconductors.

(6) Performance: Advanced functions such as arithmetic operations and data editing are achievable.

The PLC unit in a CNC system (Figure 3.15) is similar to a general PLC system but includes an auxiliary controller that assists with some functions of the NCK unit. The following functions are necessary for this setup:

(1) A dedicated circuit for communication with the NCK unit.

(2) Dual port RAM to support high-speed communication.

(3) Memory for data exchanged during high-speed communication with the NCK unit.

(4) A high-speed input module for rapid control tasks, such as turret control.

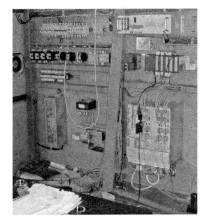

Figure 3.15 The PLC Unit in a CNC System

In practice, various PLC languages are used based on the preferences of CNC and PLC manufacturers. This diversity can lead to challenges in maintainability and users training. To overcome these issues, the standard PLC language (IEC-1131-3) was established and has gained widespread usage. The standard, IEC-1131-3, defines five kinds of programming languages:

(1) Structured text (ST).

(2) Function block diagram (FBD).

(3) Sequential function chart (SFC).

(4) Ladder diagram (LD).

(5) Instruction list (IL).

It is now necessary for users to edit programs based on the standard language, and developers are required to implement applications for interpreting and executing PLC programs.

A PLC is a digital computer used for automating electromechanical processes, such as controlling machinery on factory assembly lines, amusement rides, or lighting fixtures.

PLCs are widely used in industries and machine tools, including CNC machine tools.

As shown in Figure 3.16, the basic ladder logic symbols include:

(1) ⊣├ Normally open (NO) contact: A Normally Open (NO) contact is a type of electrical contact that remains in the open position (disconnected) when the controlling coil (or relay) is de-energized. When the coil is energized (powered), the contact closes, allowing electrical current to flow through.

(2) ⊣╱├ Normally closed (NC) contact: A Normally Closed (NC) contact is a type of electrical contact that remains closed (connected) when the controlling coil (or relay) is de-energized (off). When the coil is energized (powered), the contact opens, interrupting the flow of electricity.

(3) ◯ Output (or coil): If any left-to-right path of contacts passes electricity, the output is energized. If there is no continuous left-to-right path of contacts passing electricity, the output is de-energized.

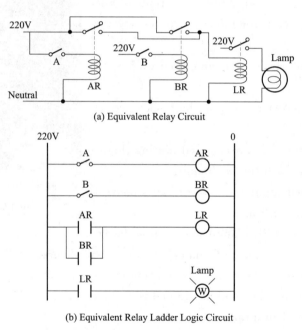

Figure 3 16   Equivalent Relay Circuit and Equivalent Relay Ladder Logic Circuit

In this architecture, a CNC unit is divided into several functional modules. Its hardware and software are designed using a modular method, meaning each functional module is built on printed circuits of the same size. The control software for these modules is also designed modularly. Hence, customers can assemble their own CNC units by combining selected functional modules into the card racks of a motherboard. Common functional modules include CNC control module, position control card, PLC card, graph display card, and communication card.

Exercises

1. Narrate the concept of MCU.
2. What is the function of GPU?
3. What are the main functions of CNC system software?
4. What is interpolation?
5. What is the principle of point-by-point comparison method interpolation?
6. What is tool radius compensation? What are the advantages of the tool redins compensation?

Chapter 4
CNC Machine Tool's Servo Systems and Position Measuring Devices

Objectives

- To understand the introduction to CNC machine tool's servo systems.
- To understand the servo motors.
- To master the position measuring devices.

4.1 Introduction of CNC Machine Tool's Servo Systems

4.1.1 Concept of Servo System

The servo system is a key subsystem of the CNC system. If we imagine that the CNC system is the "brain" of a machine tool that issues "orders", then the servo system can be considered as the "arms and legs" of the machine tool that carry out these "orders". The servo system is an important component of a CNC machine tool, and its performance directly affects the precision, speed, and reliability of the CNC machine tool.

The servo system of a typical three-axis CNC machine tool is shown in Figure 4.1. The CNC machine tool consists of mechanical components, power electronics, and CNC unit. The mechanical components mainly include spindle assembly and feed drive mechanisms. The spindle and feed drive motors, their servo amplifiers, the power supply unit, and limit switches are parts of the power electronics. The CNC unit is composed of a computer system, position and velocity sensors for each drive mechanism.

The servo system, introduced in this chapter, encompasses mechanical components, power electronics, and part of the CNC unit. In the servo system, numerical commands from the CNC machine tool are processed and amplified to the high voltage levels required

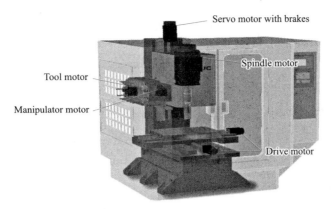

Figure 4.1 The Servo System of a Typical Three-axis CNC Machine Tool

by the drive motors. As the drives move, sensors measure the velocity and position. The CNC unit periodically executes digital control laws that maintain the feed and tool path at programmed rates by using feedback measurements from these sensors.

4.1.2 Requirements for Servo System

1. Position Accuracy

For accurate machining or other processes performed by a CNC system, the positioning system must possess a high degree of position accuracy.

Position accuracy is defined under the worst conditions, where the desired target point lies between two adjacent addressable points. Since the table can only be moved to one of these addressable points, there will be an error in the final position of the worktable.

The maximum possible positioning error occurs when the target is closer to one of the addressable points; in this case, the table would be moved to the closer control point, resulting in a smaller error. Therefore, it is appropriate to define the position accuracy based on this worst-case scenario.

2. Repeatability

Repeatability refers to the capability of the positioning system to return to a specific addressable point that has been previously programmed.

This capability can be measured in terms of the location error encountered when the system attempts to position itself at the addressable point. The location error is a manifestation of the mechanical error of a positioning system, which typically follows a normal distribution.

3. Rapid Response and Stability

Rapid response refers to the ability of the system to quickly follow instruction pulses, allowing it to play, stop, and reverse frequently. It enables the system to eliminate the load disturbance as quickly as possible. Rapid response is an important indicator of the dynamic quality of the servo system. It reflects the system's tracking accuracy.

Stability refers to the ability of the output of a serve system to oscillate as little as possible around the required output. This means the system can reach a new equilibrium state or return to its original state after experiencing a given input or external disturbance, due to the adjustment made during the short-term process. The stability of the system has a direct impact on CNC machining accuracy and surface roughness.

4. A Wide Speed Range

The speed range is defined as the ratio of maximum speed to minimum speed. It is essential that this ratio exceeds 10000∶1 while maintaining stability.

5. Large Torque at Low Speed

The servo control of feed coordinates operates under constant torque control. Large torque should be maintained across the entire speed range.

6. No Cumulative Error

Cumulative error is an error that gradually increases in degree during a series of measurements or calculations. Ensuring that there is no cumulative error is crucial for maintaining the precision and accuracy of the CNC system.

4.2　Servo Control

4.2.1　Charicteristics of Servo Motors

Machine tools, such as turning machine tools and machining centers, require high torque for heavy cutting in the low-speed range and high speed for rapid movement in the high-speed range. Also, motors with low inertia and high responsiveness are essential for machine tools that frequently perform tasks with very short machining times, such as punch presses and high-speed tapping machine tools. The characteristics required for servo motors in machine tools are as follows:

(1) Adequate power output according to the workload.

(2) Quick response to instructions.

(3) Good acceleration and deceleration properties.

(4) Broad velocity range.

(5) Safe control velocity across all speed ranges.

(6) Continuous operation over extended periods.

(7) Frequent acceleration and deceleration.

(8) High resolution to generate adequate torque, especially for small blocks.

(9) Ease of rotation and high rotational accuracy.

(10) Adequate torque generation for stopping effectively.

(11) High reliability and longevity.

(12) Ease of maintenance.

Servo motors are designed to possess the characteristics mentioned above. They comprise DC servo motors, synchronous AC servo motors, and induction AC servo motors, as shown in Figure 4.2.

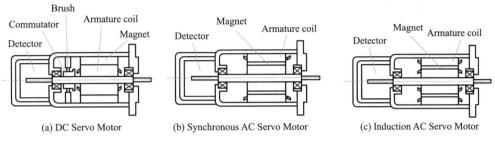

Figure 4.2　Servo Motors

The driving system is an important component of a CNC machine tool, as the accuracy and repeatability of the machining process heavily depend on the characteristics and performance of this system. The driving system is required to respond accurately to programmed instructions. Typically, electric motors are used for this purpose, although hydraulic motors may be employed for large machine tools. The motor is coupled either directly or through a gearbox to the machine lead screw, which moves the machine slide or spindle. Three types of electrical motors are commonly used in CNC applications: stepping motors, DC servo motors, and AC servo motors.

1. Stepping Motor

A stepping motor is a device that converts electrical pulses into discrete mechanical rotational motions of the motor shaft. It is one of the simplest devices applicable to CNC machine tools since it can directly convert digital data into actual mechanical displacement. Notably, it does not require any analog-to-digital converters or feedback devices for control systems, making it ideally suited for open-loop systems.

However, stepping motors are not commonly used in machine tools due to several drawbacks, including slow speed, low torque, low resolution, and a tendency to slip under overload conditions. Examples of stepping motor applications include the magnetic heads of floppy disc drives and hard disc drives in computers, daisy-wheel printers, and CNC EDM wire-cut machine tools.

There are four main types of stepping motors:

(1) Permanent magnet stepper: This can be subdivided into "tin can" and "hybrid" types, where "tin-can" is a more economical option and "hybrid" features higher quality bearings, smaller step angles, and higher power density.

(2) Hybrid synchronous stepper.

(3) Variable reluctance stepper.

(4) Lavet-type stepping motor.

A simplified unipolar stepper motor is shown in Figure 4.3.

1~4—Electromagnet.

Figure 4.3 A Simplified Unipolar Stepper Motor

Frame a—The top electromagnet 1 is turned on, attracting the nearest teeth of the gear-shaped iron rotor. With the teeth aligned to electromagnet 1, they will be slightly offset from the right electromagnet 2.

Frame b—The top electromagnet 1 is turned off, and the right electromagnet 2 is energized, pulling the teeth into alignment with it. This results in a rotation of 3.6° in this example.

Frame c—The bottom electromagnet 3 is energized, causing another rotation of 3.6°.

Frame d—The left electromagnet 4 is energized, resulting in an additional rotation of 3.6°.

When the top electromagnet 1 is enabled again, the rotor will have rotated by one tooth position. Since there are 25 teeth on the rotor, it will take 100 steps to achieve a full rotation in this example.

Generally step angle:

$$\theta = \frac{360}{mzk} \tag{4.1}$$

【拓展视频】

Where, m—Phase winding;

z—Rotor teeth;

k—Single beat $k=1$, double beat $k=2$.

CNC device: Sent out command pulses according to requirements. The number of command pulses represents the movement distance. By changing the pulse frequency, the speed can be adjusted. Each pulse causes the motor to rotate by a certain angle.

Ring distribution: Based on the direction of command instruction, the stepper motor advances one step per phase according to the power supplied. Ring distribution can be implemented using either hardware or software.

Amplifier: The amplifier enhances the ring distribution instructions for each phase, producing the drive current necessary each phase of the stepping motor.

The work principle of a stepping motor is as follows:

A stepping motor is controlled by converting logic pulses into sequetial power signals sent to the stepping motor's windings. Generally, each pulse corresponds to one rotational step of the motor. This precise control is provided by a stepping driver, which controls the speed and positioning of the motor. The stepping motor moves incrementally with each control pulse, translating digital information into exact incremental rotations without the need for feedback devices such as tachometers or encoders. Since stepping motors and their drivers operate in an open-loop control system; issues related to feedback loop phase shifts and instability, common in servo motor systems, are avoided.

With appropriate logic pulses, stepping motors can operate bi-directionally and synchronously, offering rapid acceleration, start/stop functions, and easy interfacing with other digital mechanisms. Stepping motors are characterized by low rotor moment of inertia, no drift, and noncumulative positioning errors, making them a cost-effective solution for many motion control applications. Generally, stepping motors are operated without feedback in an open-loop manner and can sometimes match the performance of more expensive DC servo systems. The only inaccuracy associated with stepping motors is a noncumulative positioning error, typically measured as a percentage of the step angle, with most stepping motors manufactured to within a 3% to 5% step accuracy.

The main difference between stepping motors and servo motors lies in the type of motor used and the control method. Stepping motors use brushless motors with 50 to 100 poles, which allows for accurate movement between step positions because of the high number of poles. Stepping motors move incrementally using pulses of current and do not require a closed-loop feedback system. In contrast, servo motors requires a feedback system to calculate the necessary current to move motors.

The performance difference between stepping motors and servo motors is primarily due to motor design. Stepping motors have significantly more poles than servo motors. As a result, a full revolution of a stepping motor requires more current pulses through its windings compared to a servo motor. Therefore, the torque of a stepping motor is greatly reduced at higher speeds compared to a servo motor. However, the higher number of poles in stepping motors provides more torques at lower speeds compared to a similar sized servo motor. The torque reduction of a stepping motor at higher speeds can be mitigated by increasing the driving voltage to the motor. Table 4-1 outlines the advantages and disadvantages of both servo motors and stepping motors.

Table 4-1 The Advantages and Disadvantages of Both Servo Motors and Stepping Motors

Items	Servo Motors	Stepping Motors
Advantages	High torque at higher speeds	Reduced cost in drive electronics
	Reduced heat production	High torque at lower speeds
		Position accuracy and repeatability
		Position stability
		High holding torque
		Easier to maintain and reliable
		Flexibility, may be used with open-loop or closed-loop configurations
Disadvantages	Drive electronics are complicated and expensive	Generating considerable heat if not using feedback
	Low torque at lower speeds	Low torque at higher speeds
		Resonance issues that must be overcome

Main control features of a stepping motor are as follows:

(1) Step angle (θ) and step error.

The step angle is the angle by which the rotor turns between two adjacent pulses. Generally, the smaller the step angle, the higher the control precision.

The step error directly affects the positioning accuracy of the components being controlled. When a single phase of the stepping motor is powered, the step error depends on the precision of the stator's sub-tooth alignment and the accuracy of the misalignment angle of each phase stator.

When multi phases of the stepping motor are powered simultaneously, the step error is influenced not only by the factors mentioned above but also by the phase current and the performance of the magnetic circuit.

(2) The maximum starting frequency.

The maximum starting frequency refers to the highest frequency at which the motor can start from a stationary position and immediately transition to steady-speed operation without losing step, under idle load conditions. If the starting frequency exceeds this limit, the stepping motor will work correctly. The maximum starting frequency is influenced by the inertia of the load on the stepping motor.

(3) The maximum operating frequency.

The maximum operating frequency is the highest frequency at which the stepping motor can run continuously without losing steps as the operating frequency gradually increases. This frequency is also load-dependent. Typically, the maximum operating frequency is much higher than the maximum starting frequency for the same load.

(4) Torque-frequency characteristic.

In continuous operation mode, the electromagnetic torque of the stepping motor declines sharply as the operating frequency increases. The relationship between torque and frequency is known as the torque-frequency characteristic.

There are several important criteria involved when selecting an appropriate stepping motor:

(1) Desired mechanical motion.

(2) Required speed.

(3) Load.

(4) Stepping mode.

(5) Winding configuration.

【拓展图文】

Motion requirements, load characteristics, coupling techniques, and electrical needs must be understood before the system designer selects the optimal combination of stepping motor, driver, and controller for a specific application. These key factors are crucial when determining an optimal stepping motor solution. The system designer should adjust the characteristics of the elements under their control to meet application requirements. Considerations should include the selection of the stepping motor, driver, and power supply, as well as mechanical transmission options, such as gearing and load weight reduction through the use of alternative materials.

Most stepping motors are labeled with key specifications, as shown in Figure 4.4. The major points on the label typically include the voltage, resistance, and the number of degrees per step. Understanding the number of degrees per step is vital for configuring the software to control the machine accurately later on. For a three-axis machine, it is highly recommended that both the X and Y axes use identical motors. While it is not catastrophic if the motors do not match, it can create complications and make the setup and operation more challenging down the vine.

Figure 4.4 A Label of A Stepping Motor

2. DC Servo Motor

The DC servo motor is the most common type of feed motors used in CNC machine tools. Its operation is based on the rotation of an armature winding within a permanently energized magnetic field. The armature winding is connected to a commutator, which is a cylinder made of insulated copper segments mounted on the motor shaft. DC current is supplied to the commutator through carbon brushes, which are connected to the machine terminals. Motor speed is adjusted by varying the armature voltage, while motor torque is controlled by managing the armature current. To achieve the necessary dynamic performance, DC servo motors operate within a closed-loop system, equipped with sensors that provide the velocity and position feedback.

The construction of a DC servo motor is shown in Figure 4.2 (a). The stator consists of a cylindrical frame that serves as both a passage for magnetic flux and mechanical supporter, with a magnet attached to the inside of the frame. The rotor comprises a shaft and a brush assembly. The commutator and rotor metal supporting frame (rotor core) are mounted on the outside of the shaft, with the armature coil wound around the rotor core. A brush that supplies current through the commutator is integrated with the armature coil. Additionally, a detector for measuring rotation speed, typically an optical encoder or tachogenerator, is built into the back of the shaft.

In a DC servo motor, a controller can be easily designed by using a simple circuit because the torque is directly proportional to the current. However, the motor's output power is limited by the heat generated inside the motor due to current flow. Therefore, efficient removal of the heat is essential to generate high torque. DC servo motors have a broad velocity range and are relatively inexpensive. However, friction with the brushes results in mechanical loss and noise, and it is required to regularly maintain the brushes.

3. AC Servo Motor

In an AC servo motor (Figure 4.5), the rotor is a permanent magnet, while the stator is equipped with three-phase windings. The speed of the rotor is directly related to the rotational frequency of the power supply.

Figure 4.5　An AC Servo Motor

AC servo motors are gradually replacing DC servo motors, primarily because they do not have commutators or brushes, resulting in virtually no maintenance requirements. Additionally, AC servo motors offer a smaller power-to-weight ratio and faster response times.

The structure of an AC servo motor is shown in Figure 4.6.

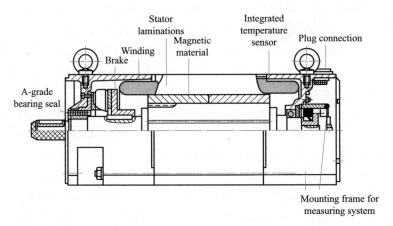

Figure 4.6　The Structure of an AC Servo Motor

(1) Synchronous AC servo motor.

The stator consists of a cylindrical frame and a stator core, which is located inside the frame. An armature coil is wound around the stator core, and the ends of the coil are connected to lead wires that supply current. The rotor consists of a shaft and a permanent magnet, and the permanent magnet is attached to the outside of the shaft.

In a synchronous AC servo motor, the permanent magnet is mounted on the rotor, and the armature coil is wound around the stator, unlike in a DC servo motor. This design allows for current supply from the outside without requiring a commutator, leading to the term "brushless servo motor". This structure makes it possible to cool down a stator core directly from the outside, effectively preventing temperature from increasing. Furthermore, synchronous AC servo motors do not have limitations on maximum velocity due to rectification spark, allowing for excellent torque performance in high-speed applications. In addition, because there are no brushes, these motors can operate for long periods without maintenance.

Like DC servo motors, synchronous AC servo motors use optical encoders or resolvers to detect rotational velocity. They typically employ ferrite magnets or rare earth magnets for the rotor. In this type of AC servo motor, the armature's contribution is linearly proportional to the torque, facilitating easy stopping and effective dynamic braking during emergency stops.

However, as a permanent magnet is used, the structure is more complex, and rotor position detection is needed. Additionally, the current from the armature can include high-frequency components, which may lead to torque ripple and vibrations.

(2) Induction AC servo motor.

The structure of an induction AC servo motor is similar to that of a standard induction motor. When multiphase alternating current flows through the stator coil, it induces a current in the rotor coil, generating torque. The stator consists of a frame, a stator core, an armature coil, and lead wires, while the rotor comprises a shaft and a rotor core made of conductive materials.

Induction AC servo motors have a simple structure and do not require a detector for the relative position between the rotor and the stator. However, because the field current needs to flow continuously during stopping, this can lead heat loss, making dynamic braking impossible.

The strengths, weaknesses and characteristics of servo motors mentioned above are summarized in Table 4-2.

Table 4-2 The Summary of Servo Motors

Items	DC Servo Motor	Synchronous AC Servo Motor	Induction AC Servo Motor
Strengths	Low price, Broad velocity range, Easy control	Brushles, Easy stop	Simple structure, No detector needed
Weaknesses	Heat Brush wear, Noise, Position-detection needed	Complex structure, Torque ripple, Vibration, Position-detection needed	Dynamic braking impossible, Heat loss
Capacity	Small	Small or medium	Medium or large

Continued

Items	DC Servo Motor	Synchronous AC Servo Motor	Induction AC Servo Motor
Sensor	Unnecessary	Encoders, Resolvers	Unnecessary
Life length	Depends on brush life	Depends on bearing life	Depends on bearing life
High speed	Inadequate	Applicable	Optimized
Resistance	Poor	Good	Good
Permanent magnet	Exists	Exists	Does not exist

4. Linear Motor

A linear motor (Figure 4.7) is essentially an AC rotary motor that has been laid out flat. The principle of producing torque in rotary motors is adapted to generate force in linear motors.

Figure 4.7 A Linear Motor

By leveraging the electromagnetic interaction between a coil assembly and a permanent magnet assembly, electrical energy is converted into linear mechanical energy, resulting in linear motion. Since the motion is linear rather than rotational, it is referred to as a linear motor.

Linear motors offer several advantages, including high speeds, precision, and fast response times. In the 1980s, machine tool builders started incorporating linear motors into their designs, using common motion control servo drives.

Among the different designs of linear motors, permanent magnet brushless motors stand out due to their high force density, high maximum speed, and stable force constant. The lack of a brushed commutator assembly offers additional benefits, such as reduced maintenance, increased reliability, and smoother operation.

An iron-core brushless linear motor [Figure 4.8(a)] is similar to a conventional brushless rotary motor that has been split axially and laid out flat. In this configuration, the "unrolled" rotor is a stationary-plate consisting of magnets affixed to an iron backplate, which the "unrolled" stator is a moving coil assembly made up of coils wound around an iron core. The coil windings are typically arranged in a conventional three-phase configuration, with commutation often performed by Hall-effect sensors or sinusoidal control. This design is highly efficient and well-suited for generating continuous forces.

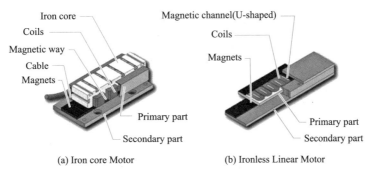

(a) Iron core Motor   (b) Ironless Linear Motor

Figure 4.8 Linear Motors

An ironless linear motor [Figure 4.8(b)] features a stationary U-shaped channel filled with permanent magnets arranged along both interior walls. A moving coil assembly traverses between two opposing rows of magnets. Commutation is handled electronically, either via Hall-effect sensors or sinusoidal control. Ironless linear motors offer several advantages, including lower core mass, reduced inductance, and smooth motion without cogging, as there is no attractive force between the frameless components.

4.2.2 Position Control

1. Speed Control

Motor speed is measured and compared with a reference signal. Usually, the speed is determined by calculating the time derivative of the rotor angle. These signals are then forwarded to the speed control system, which produces a reference value. In combination with other relevant signals from various parts of the drive system, these inputs are employed in the flux and torque control system to regulate motor performance.

2. Velocity Feedback Device

The actual speed of the motor can be measured by the voltage generated from a tachometer mounted at the end of the motor shaft. A DC tachometer functions as a small generator that produces an output voltage proportional to the motor's speed. The voltage generated is compared to the command voltage, which corresponds to the desired speed. The difference between these voltages can be used to adjust the motor, reducing any speed errors.

Figure 4.9 A Tacho-generator

A tacho-generator (Figure 4.9) is AC or DC generator that outputs a voltage proportion to the rotational speed of a shaft in a rotating electrical machine (such as an electric motor). They are used to measure both the speed and the direction of rotation.

3. Position Control

Position control requires a high level of operating precision. In this mode, the motor starts, runs, and stops after rotating through a specific angle. Position control systems are widely used in manufacturing industries, for applications like elevators, material handling equipment, packaging systems, and processing lines, as well as in machine tools.

An angular position sensor, usually an encoder, provides the necessary control feedback. In position control systems, the position controller is the main outer loop, with speed and torque controllers functioning as inner ones. The position controller can be implemented using various control strategies, including linear controllers (like PI or PID), fuzzy or neural-fuzzy systems, or variable structure control types.

4.3 Position Measuring Devices

In a closed-loop system, information about the output is fed back and compared with the input. Thus, position measuring devices are required to facilitate this feedback loop. These devices are typically either rotary or linear, providing both position and velocity signals.

For a CNC machine tool to operate accurately, the positional values and speeds of the machine tool's axes need to be constantly updated. Two main types of feedback devices are used: position measuring devices and velocity measuring devices. Here we only introduce position measuring devices. Position measuring devices can be classified into two types: linear transducers for direct positional measurement and rotary encoders for angular or indirect linear measurement.

A linear transducer is a device mounted on the machine table to measure the actual displacement of the slide. By doing so, it mitigates errors from backlash in screws and motors, ensuring more accurate feedback data. Linear transducers are generally considered to provide higher accuracy but more expensive than other position measuring devices that rely on screws or motors.

4.3.1 Requirements for Position Measuring Devices

To effectively determine the slideway position of each axis in a CNC machine tool, a position measuring device is required. This device continuously monitors and compares the present position with the command position during axis movement.

The following requirements should be considered when selecting position measuring devices for CNC machine tools:

(1) Reliability: The devices should exhibit maximum reliability and minimum interference with the quantity being measured.

(2) Accuracy and speed: The devices must satisfy high accuracy across a range of speeds.

(3) Ease of installation and maintenance: The devices should be easy to install and maintain, and it should be adaptable to the working environment.

(4) Cost-effectiveness: The purchase cost of the devices should be low.

4.3.2 Classifications for Position Measuring Systems

Position measuring systems can be classified according to the following considerations:

(1) Direct and indirect measuring systems.

In a direct measuring system, the position measuring devices are directly coupled to the machine tool's slideway or table. This setup is independent of the lead screw and the drive element, thereby eliminating errors associated with those components. However, direct measuring system tends to be more costly because the accuracy must be maintained throughout the measuring length. Additionally, it can introduce the full mechanical resonance of the machine tool into the feedback loop. Direct measurement can be achieved with linear measuring devices such as optical gratings or inductosyns. Despite their higher costs, the demand for accurate components is increasing.

In an indirect measuring system, the position measuring devices mounted on the lead screw and the drive element of the machine member being positioned. Indirect measurement has advantages, including lower cost, simpler mounting on the machine tool, and easier maintenance. However, the accuracy may be compromised by factors such as backlash and pitch errors in the lead screw. Indirect measuring systems usually use rotary position measuring devices, such as encoders and resolvers.

(2) Incremental and absolute measuring systems.

In an incremental measuring system, each displacement is determined by measuring the change from the preceding position. While this method is straightforward, it has notable disadvantages, such as the inability to determine the absolute position without reference to previous measurements. This can lead to cumulative errors, where an error in a prior measurement affects all subsequent readings.

In contrast, an absolute measuring system determines all coordinates from a fixed datum position or the center of coordinates without reference to previous positions. This approach simplifies restarting after a power failure (midway starts) and reduces cumulative errors. However, digital absolute measuring devices can be expensive especially at high resolutions.

(3) Digital and analogue measuring systems.

In a digital measuring system, position or displacement is measured using discrete values. Digital measuring systems can be incremental (e.g., optical gratings and incremental encoders) or absolute (e.g., absolute encoders). Digital signals are easier to store, more reliable to transmit, and error-free in reproduction. Additionally, digital measuring systems do not require a separate analog-to-digital converter.

In an analogue measuring system, displacement is converted into a continuous physical representation that can be easily measured. The physical analogs can vary widely. A common example is an inductive transducer, whose output is continuously proportional to displacement (e.g., resolvers and inductosyns).

In an analogue system, a signal such as electrical voltage represents a physical position, meaning that a specific slideway displacement correlates with an induced voltage. Analogue measuring devices are generally simpler, more robust, and less expensive than their digital counterparts. However, because they provide continuous measurements, displaying the position of the moving members can be more challenging. Analog displacemen tmeasuring devices are typically suitable for small and medium-sized CNC machine tools.

4.3.3 Common Position Measuring Devices

【拓展视频】

1. Resolvers

A resolver (Figure 4.10) is a position sensor or a transducer that measures the instantaneous angular position of the rotating shaft to which it is attached. Resolvers, along with their close relatives known as synchros, have been in use since World War Ⅱ for military applications, such as measuring and controlling the angle of gun turrets on tanks and warships.

Resolvers are typically constructed like small motors, consisting of a rotor (which is attached to the shaft whose position is to be measured) and a stator (the stationary part that produces the output signals). All resolvers produce signals that are proportional to the sine and cosine of their rotor angles. Since every angle corresponds to a unique combination of sine and cosine values, a resolver provides absolute position information within one complete revolution (360°) of its rotor. This absolute position capability, as opposed to incremental positioning, is one of the main advantages of resolvers over incremental encoders.

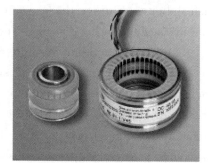

Figure 4.10 A Resolver

2. Optical Gratings

An optical grating (Figure 4.11) consists of two transparent scales made from glass, which require delicate handling. These scales, created through photo etching or vapor deposition, have a series of parallel lines that are closely and uniformly spaced. The transparent and opaque (black) lines can be of referred to as the equal width. Typically, one scale is longer and is referred to as the "scale grating", which is mounted on the moving member. The shorter scale is called the "index grating" and is mounted on the stationary member.

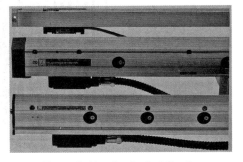

Figure 4.11　An Optical Grating

As shown in Figure 4.12, an optical grating measuring system consists of a light source, a collimating lens, a scale grating, and a grating pitch, etc. The long, fixed scale grating extends over the entire length of the machine tool's slideway travel, with the short index grating positioned over it. All except for the scale grating are housed in a reading head.

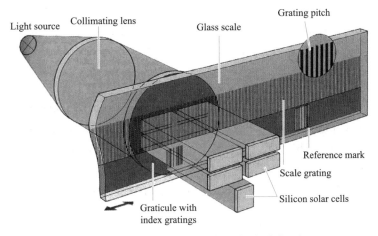

Figure 4.12　The Structure of an Optical Grating

The principle behind the optical grating is based on the well-known "Moire fringe effect". The lines of the two scales are slightly tilted relative to each other by a very small angle; θ. This slight misalignment causes an interference effect at the intersection of the lines, creating a pattern known as a Moire fringe.

In practice, when the slideway moves, the lines on the two scales become displaced relative to each other. This displacement causes the fringe pattern to travel perpendicularly across the scales, with the direction of movement dependent on whether the slideway moves to the right (positive) or to the left (negative). As shown in Figure 4.13, a dark or light fringe pattern is produced across the width of the grating and moves in the sequence illustrated.

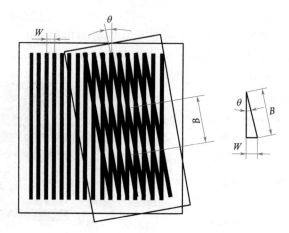

W—Width of Moire fringe; B—Grid distance.

Figure 4.13 Moire Fringe Effect

$$W = \frac{B}{\sin\theta} \tag{4.2}$$

θ is very small, $\sin\theta \approx \theta$.

$$W = \frac{B}{\theta} \tag{4.3}$$

Suppose $B = 0.01 \text{mm}$, $\theta = 0.01$ radian, then $W = 1 \text{mm}$. It means the magnification factor is 100.

3. Encoders

Encoders (Figure 4.14) are devices used to measure angular position or displacement. They can be classified into two primary types:

(1) Absolute encoders produce a code value that directly represents the absolute position.

(2) Incremental encoders generate digital signals that increase or decrease the measured value in incremental steps.

A rotary encoder is typically mounted at the end of a motor shaft or screw to measure the angular displacement. However, it cannot measure linear displacement directly, which can introduce errors due to factors like backlash in the screw or the motor.

These errors are generally compensated by the machine tool builder during the calibration process.

An absolute encoder (Figure 4.15) is a kind of rotary measurement devices that immediately determine the actual position value when the system is switched on. Absolute encoders do not require a counter, as the measured value is directly derived from the graduation pattern. The output from absolute encoders is typically in the form of either pure binary code or Gray code.

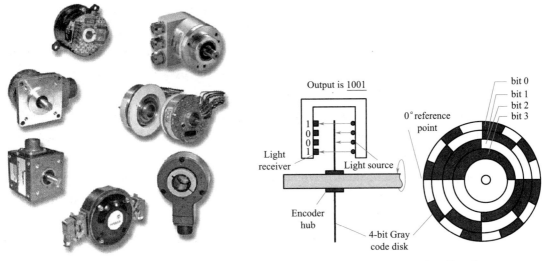

Figure 4.14　Encoders　　　　　Figure 4.15　An Absolute Encoder

Absolute encoders are available in three types: brush, optical, and magnetic.

Unlike absolute encoders, incremental encoders act as pulse generators or counters. Although incremental encoders are the simplest form of position measuring devices, they do not provide an unambiguous indication of position like absolute encoders. Instead, they provide relative information, and the output impulses must be counted to determine position.

As shown in Figure 4.16, an incremental encoder consists of a transparent disk mounted on a shaft. This disk has a precise circular pattern of alternating clear and opaque segments around its periphery. A fixed light source is provided on the one side of the disk, while a sillicon solar cell is placed on the other side.

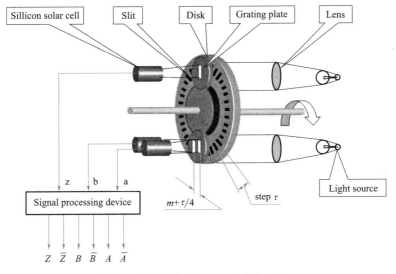

Figure 4.16　An Incremental Encoder

As the disk rotates, the light intermittently falls on the sillicon solar cell, producing a sinusoidal output signal with slow rising and falling edges in the millivolt range. This sinusoidal signal is then amplified and passed through a shaping circuit, which converts it into a square wave. The square wave is further processed by a differentiating element to generate short-duration pulses. A pulse produced each time a line on the disk interrupts the light path. The number of pulses generated per revolution of the disk is directly related to the number of lines etched on the disk, known as resolution. A typical encoder disk may have anywhere from 200 to 18000 lines.

Incremental encoders are shaft-driven devices that output electrical pulses at their terminals. The number of pulses generated determines the measured decides distance. The pulse count can be used to accurately determine the linear axis position of the machine tool's workable by factoring in the lead screw pitch and any gear ratios.

Exercises

1. State the composition of servo motors.
2. What are the characteristics of servo motors?
3. What is the work principle of open-loop control of stepping motors?
4. How to select stepping motors?
5. What is the work principle of AC servo motors?
6. Narrate the work principle of optical gratings.
7. Narrate the work principle of incremental encoders.

Chapter 5
Mechanical Construction and Tool System of CNC Machines

Objectives

- To understand the CNC machine tools and the CNC machining centers.
- To understand the mechanical system of CNC machine tools.
- To understand the tool system of CNC machine tools.

5.1 CNC Machine Tools and CNC Machining Centers

The purpose of a machine tool is to cut away surplus material. Usually, the materials are supplied to create a workpiece of the required shape and size, achieving an acceptable level of accuracy and surface finish. CNC machine tools must possess certain capabilities to fulfill these requirements:

(1) Able to hold the workpiece and cutting securely.

(2) Endowed the sufficient power to enable the tool to cut the workpiece material at economical rates.

(3) Capable of displacing the tool and workpiece relative to one another to produce the desired shape. The displacements must be controlled with a degree of precision that ensures the desired accuracy of surface finish and size.

【拓展视频】

Early CNC machine tools were designed so that the operator could stand in front of the CNC machine tool while operating the controls. This design is no longer necessary, as operators do not directly control CNC machine tool movements. On conventional machine tools, only about 20% of the time was spent removing materials. With the addition of electronic controls, the actual time spent removing metal has increased to 80% or higher. This advancement has also reduced the time required to bring tools into each machining position.

5.1.1 CNC Lathe

The engine lathe, one of the most productive machine tools, has always been an efficient means of producing round parts. Features of a CNC Lathe include:

(1) The tool or material moves during operation.

(2) Tools can operate in 1 to 5 axes.

(3) Larger CNC machine tools are equipped with MCUs that manage operations.

(4) Movement is controlled by motors.

(5) Feedback is provided by sensors.

(6) Tool magazines are used to change tools automatically.

Tools:

(1) Most tools are made from high speed steel (HSS), tungsten carbide, or ceramics.

(2) Tools are designed to direct waste material away from the workpiece.

(3) Some tools require coolant, such as oil, to protect the tools during operation.

Tool paths, cutting, and plotting motions:

(1) Tool paths describe the route the tool takes during operation.

(2) Motion can be described as point-to-point, straight cutting or contouring.

(3) Speeds refer to the rate at which the tool operates, such as revolutions per minute (r/min).

(4) Feeds refer to the rate at which the tool and the workpiece move in relation to each other.

(5) Feeds and speeds are determined by factors such as cutting depth, material, and quality of finish required. For example, harder materials require slower feeds and speeds.

(6) Rouging cuts remove larger amounts of material compared to finishing cuts.

(7) Rapid traversing allows the tool or the workpiece to move quickly when no machining is taking place.

1. Purpose of CNC Lathe

Both CNC lathes and conventional lathes are mainly used for machining revolving body parts, such as axes and plates. However, compared to conventional lathes, CNC lathes offer hgher machining accuracy, more consistent machining quality, and greater efficiency. They are also more versatile and require less manual labor. Especially, CNC lathes are well-suited for machining complex-shaped parts.

2. Conventional Cutters and Fixtures of CNC Lathe

The turning tools used in CNC lathes are similar to those found in conventional lathes.

They mainly include two modes: welding mode and mechanically clamped mode. In CNC turning operations, typical tools include small radius circular turning tools, non-rectangular slotting tools, and thread turning tools. However, these specialized tools are not commonly used in practical applications and may only be employed occasionally. Necessary detailed instructions should be provided in the technological documentation and machining program list.

Regarding the fixtures for CNC lathes, both general three-jaw and four-jaw centering chucks are usually used, along with automated hydraulic, electric, and pneumatic fixtures, especially in mass producing settings. In addition, other appropriate fixtures are usually applied for machining both axes and plate parts.

(1) Fixtures for axes.

Fixtures for axes typically include automatic clamp chucks, centers, three-jaw chucks, and rapidly adjustable universal chucks. When machining axes on a CNC lathe, the workpiece is hold firmly between the spindle center and the tailstock center, with the tailstock center promoting rotation around the spindle. These fixtures are designed to transmit sufficient torque during round turning to accommodate the rapid rotational speed of the spindle.

(2) Fixtures for plate parts.

Fixtures used for machining plate parts include adjustable dog chuck and rapidly adjustable chuck. These fixtures are specifically designed for CNC lathes that do not utilize a tailstock chuck, ensuring efficient and precise machining of plate parts.

3. CNC Lathe Tool Compensation

When machining a workpiece on a CNC lathe, multiple tools are often required, each with a different nose location due to varying shapes and sizes. Tool compensation in CNC lathes is a method to account for these positional differences of the tools noses. The goal is to align the tool nose of each tool to a common clamped location within the same work coordinate system. This ensures that all tool noses move consistently within the predefined coordinates of the workpiece, allowing for the accurate machining. Here are some common tool compensation methods used in CNC lathes:

(1) Automatic tool compensation: It simplifies the setup process by automating the compensation adjustments.

(2) Trial cutting: It is mainly used for the CNC lathes with either closed-loop or open-loop control systems.

(3) Tool compensation inside the machine tool: It uses a fixed touch inside the machine tool to make contact with the tool. The machine tool measures the tool deflection from its interred path and corrects any discrepancies.

(4) Tool compensation by reference point location: To ensure a proper tool positioning, adjust the mechanical stopper location of each coordinate axis on the machine bed

using the CNC system parameters. Set the reference points for tool compensation to correspond with the tool's starting points. This setup allows the machine bed to return to the reference point of the staring operator, ensuring the tool nose return to its original starting position.

4. Components of a CNC Lathe

A CNC lathe is a machine tool designed to remove material from a workpiece blank that is clamped on an axis and rotated about it. This machine primarily focuses on machining surfaces and end faces of rotary parts made from metals various. Typically, there operations are conducted using a sharp single-point tool. CNC lathes use a turret to hold tools rigidly and move them accurately.

In addition to holding tools, the turret features an automatic tool changing function, which allows for the quick removal of an old tool and insertion of a new tool into its cutting position. A front turret is built to move tools from below the spindle centerline up to the workpiece, while a rear turret moves cutters from above the spindle centerline down to the workpiece. This configuration enables machines equipped with both front and rear turrets to execute cutting operations from above and below the workpiece simultaneously.

The following are components of a CNC lathe (Figure 5.1):

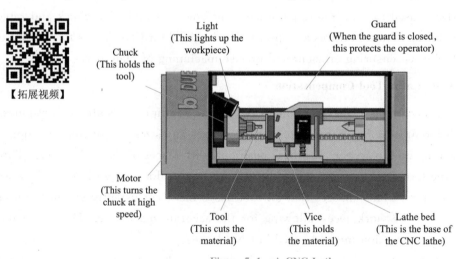

Figure 5.1 A CNC Lathe

(1) Vice.

The vice holds the material to be cut or shaped securely in place. If the material is not held firmly, it may "fly" out of the vice when the CNC begins machining. Typically, the vice functions like a clamp to ensure the material is correctly positioned.

(2) Guard.

The guard protects the operator from potential hazards. During machining, small pieces can be ejected from the material at high speed, posing a danger if they hit the operator. The guard completely encloses the dangerous areas of the CNC lathe.

(3) Chuck.

The chuck connects the spindle and clamps the workpiece. It securely holds the material to be shaped. The material must be placed carefully to prevent it from being ejected at high speed during operation of the CNC lathe when it is working.

(4) Motor.

The motor, housed within the CNC lathe, rotates the chuck at high speed, enabling the cutting process.

(5) Lathe bed.

The lathe bed serves as the base of the CNC lathe. It is typically bolted down to prevent movement caused by vibrations. The lathe bed is the main frame that supports all components and provides a path for chips to fall away easily. The slanted design of the lathe bed facilitates the easy disposal of chips, and it has guideways that allow the CNC lathe carriage to slide lengthwise easily. The height of the CNC lathe should be appropriate for the technician to work comfortably.

(6) Tool.

Tools are usually made from high-quality steel and are the components that actually cut the material to be shaped.

5.1.2 CNC Machining Centers

Milling is the most versatile machining process. Metal removal is accomplished through the relative motion of a rotating, multi-edge tool, and multi-axis movement of the workpiece. Milling is a form of interrupted cutting, where the repeated cycles of entry and exit motions of the tool facilitate actual metal removal and generate discontinuous chips. The milling machine has always been one of the most versatile machine tools used in the manufacturing industry. Operations such as milling, contouring, gear cutting, drilling, boring, and reaming are only a few of tasks that can be performed on a milling machine. A machining center is one type of milling machines.

All milling machines, from compact tabletop models to standard vertical knee mills and massive CNC machining centers, operate on the same principles and parameters. The most important features of these operating parameters are:

(1) Cutting speed: The speed at which the tool engages the workpiece.

(2) Feed rate: The distance the tool edge travels in one cutting revolution.

(3) Axial depth of cut: The distance the tool is set below an unmachined surface.

(4) Radial depth of cut: The amount of work surface engaged by the tool.

The capabilities of the milling machine are determined by motor horsepower, which influences the maximum spindle speeds and spindle taper size.

A machining center, a specialized milling machine, is designed for both milling and hole-making on a variety of non-round or prismatic shapes. A distinctive feature of the machining center is its tool changer system. This system moves tools from storage to the spindle and back in rapid sequence. While most machining centers store and handle 20 to 40 individual tools, some can manage inventories of over 200 tools.

The heart of the milling operation is the milling tool, a rotary tool with one or more cutting edges, each of which removes a small amount of material as it enters and exits the workpiece. The variety of tool types is extensive. One basic type is the face mill tool, which is used for milling flat surfaces. These tool, used at high speeds, can range in diameter from three inches to up to two feet. Some face mills are also capable of simultaneously milling a shoulder that is square to the surface.

The work that requires edge preparation, shoulders, and grooves, is accomplished using other types of cutting mills. An end mill tool is a tool with cutting edges on its end as well as on its periphery. End mills are typically used for short, shallow slots, and some edge finishing. For longer and deeper slots, circular grooving or slotting tools are more suitable, as end mills are more easily deflected during heavier cuts. Chamfers and contour milling are performed with specially shaped end mills.

In all kinds of milling, a critical component is the work-holding device, which must be capable of being quickly changed over to present a new work or work surfaces to the tool. Machining centers can utilize long machine beds, pallet changers, and multi-storied "tombstone" part holders, allowing the new work to be set up and positioned while previously set-up workpieces are being milled.

Figure 5.2 A CNC Machining Center

A CNC machining center (Figure 5.2) can incorporate two useful accessories. One is the touch-trigger probe, along with its computer software, which dimensionally checks workpiece measurements before removal from the machining center. This probe is stored with other tool for quick application. The second accessory is the tool presetting machine, which allows technicians to assemble the tool according to the programmed part requirements before placing tools in the machining center's tool storage.

Machining centers are CNC machine tools equipped with ATCs and tool magazines. They have been improved by increasing tool magazine capacity and incorporating and rotating worktable based on traditional CNC milling machines. Therefore, machining centers can perform various functions, including milling, boring, and drilling. The main characteristics of a CNC machining center is as follows:

(1) High centralized working procedure: Once a part is secured, machining process can be completed on multiple surfaces.

(2) Automatic dividing unit: It can be equipped with an automatic deviding unit (or rotating worktable) and a tool magazine system.

(3) Automatic speed and feed adjustment: The spindle speed, feed quantity, and motion path of the tool can be automatically adjusted relative to the workpiece.

(4) Increased productivity: The productivity of a CNC machining center is five or six times higher than that of a traditional CNC machine tool. It is especially suited for machining complex-shaped parts that require higher precision and frequent changes in variety.

(5) Low operator labor intensity: The operator's labor intensity is very low, though the machine structure is complex, requiring a high level of technical skill from the operator.

(6) High cost: The cost of these machines is relatively high.

The components of different kinds of CNC machining centers mainly composed of general parts, spindle parts, CNC systems, ATCs, and various accessories.

CNC Machining centers may be either vertical or horizontal, with some universal types capable of both orientations. The vertical type is often preferred when work is done on a single face. With the use of rotary tables, more than one side of a workpiece, or several workpieces, can be machined without the operator's intervention. Vertical CNC machining centers using a rotary table have four axes of motion: three lineal motions of the table and one rotary motion of the table itself.

The classification of CNC machining centers includes several types, each designed for specific applications:

(1) Vertical CNC machining center: Its spindle is vertical. It is typically used for machining plate parts. A rotation workable can be mounted on the level worktable to machine helical lines.

(2) Horizontal CNC machining center: Its spindle is horizontal, and is usually equipped with a dividing rotation table that allows for inclined movement along three to five motion coordinates. It is typically used for machining box-like parts.

(3) Planer CNC machining center: Its spindle is usually vertical, and is equipped with changeable spindle head accessories. It is typically used for machining large-sized or complex-shaped parts.

(4) Multipurpose CNC machining center: It combines both horizontal and vertical function-alities. Once a workpiece is secured, it can machine all surfaces except for external faces. The multipurpose CNC machining center can reduce configuration errors and eliminate the need for reclamping the workpiece, leading to higher productivity and lower costs.

(5) CNC Machining center with mechanics and tool magazine: It uses an ATC, commonly involving a tool magazine and mechanical systems. It offers a wide range of machining capabilities, making it very versatile.

(6) Turret magazine CNC machining center: It is widely used for mechining small-sized parts.

To obtain high accuracy and repeatability, the machine slides and the driving lead screws are of vital importance. Slides are machined to high accuracy and coated with antifriction material such as PTFE (polytetrafluoroethylene) or turcite, which help reduce stick-slip phenomena. Large diameter recirculating ball screws are employed to eliminate backlash and lost motion, which are essential for maintaining precision in CNC machining-centers.

Other design features, such as rigid and heavy machine structure, short machine table overhang, quick change tooling system, also contribute to the high accuracy and repeatability of CNC machining centrs.

5.2　Mechanical System of CNC Machine Tools

The mechanical system is composed of three main sections: the drive, the guide rail, the frame, the gantry sides, the base table, and the chip conveyor.

【拓展视频】

1. Drive

The drive of CNC machine tools converts torque provided by electric motors into the linear motion of the tool head. Screws with threaded nuts offer a simple and compact method for transmitting this power. A ball screw and ball nut system will be used because of its low friction and high efficiency. ACME (association of consulting management engineers) screws will not be used because their advantages, such as larger weight-supporting capacity and self-locking simplicity, are not applicable for this machine tool. Instead, the ability of ball screws to reduce the required torque necessary for achieving specified linear speeds, due to their superior efficiency, makes them an obvious choice for all three axes. In addition, the minimal heat generation caused by friction and increased reliability further support the decision to implement ball screws and ball nuts for power transmission.

2. Guide Rail

The guide rail supports the weight of the gantry and the tool head while ensuring proper alignment during the movement of the gantry. The linear rod guide rails are made from case-hardened steel shafts with ball bushings. However, a more complex shaft or support rail may be required if the weight and loads on the gantry create deflections that exceed the specified tolerances of the machine tool.

3. Frame

The frame is divided into the gantry sides and the base table. CNC frame materials must possess sufficient strength to support the weight of the gantry, and the tool head, as well as withstand the forces generated during the milling process.

Stiffness is also essential to prevent deflections caused by both static and dynamic forces resulting from the acceleration of the tool head. The frame's weight is important because its mass contributes to both static and acceleration forces. The ideal frame material should fulfill all three requirements (strength, stiffness, and weight), offer excellent machinability, and be available at a low cost. Opon reviewing various materials, high-density polyethylene (HDPE) emerged as the best choice due to its combination of the five selection factors. Among plastics, it is relatively inexpensive, easy to machine, and provides sufficient strength and rigidity.

4. Gantry Sides

The gantry sides support the weight of the upper gantry and the tool head while traveling on the lower guide rails. HDPE will be used to construct the gantry sides, as it is lightweight, which helps reduce inertia forces during acceleration.

5. Base Table

The base table supports the material to be worked and serves as the foundation of the CNC machine tool. Constructing the base table will require a sufficient amount of material and involve extensive machining and assembly. Due to its low cost and ease of machining, HDPE will be used for the base material. In this application, the strength and stiffness properties of HDPE will be tested. The weight supported by the lower guide rails, which allow the machine to move along the length of the table, might create excessive deflections in the sides of the base table. This could cause a displacement of the tool head that exceeds design tolerances. It is believed that HDPE will withstand these forces and maintain tolerances. If deflections become problematic, a hybrid system will be considered, utilizing aluminum to reinforce the HDPE base table.

6. Chip Conveyor

The chip conveyor (Figure 5.3) is a moveable belt designed to remove metal chips produced by CNC machine tools. In some lathes, the chip conveyor is located inside the coolant tank. Chip conveyors usually use heavy-duty steel link belts arranged in a serpentine configuration.

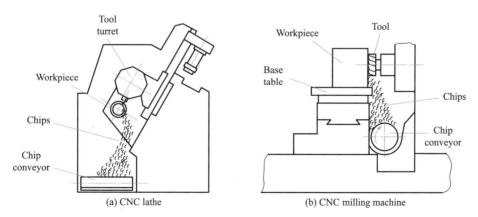

(a) CNC lathe   (b) CNC milling machine

Figure 5.3 Chip Conveyors Used in CNC Machine Tools

【拓展视频】

5.2.1 Spindle Design

The spindle is the workhorse of the CNC machining center, and its design is crucial to the CNC machining center's performance and longevity. When purchasing a CNC machining center, most people focus on the basic specifications such as maximum spindle speed, peak spindle motor horsepower, and maximum spindle motor torque.

However, it is equally important to ensure that the manufacturer has invested in high-quality components that enhance the spindle's durability and reliability (Figure 5.4).

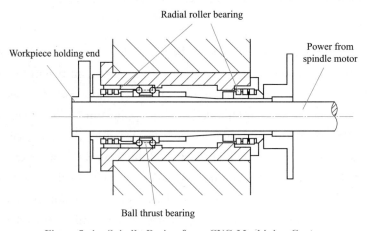

Figure 5.4 Spindle Design for a CNC Machining Center

The power required depends on the material being cut. While torque, speed, and horsepower are important to consider when purchasing a vertical CNC machining center, there are additional factors that will critically impact the overall performance of the spindle and the satisfaction with the investment.

1. The Importance of Spindle

At first glance, it might appear that the spindle's role in a vertical CNC machining center is straightforward—the tool cuts the metal, the table moves, the motion control system ensures precision, and the software handles the rest. The spindle seems to be just a motor that holds a tool and follows commands from a servo.

While this depiction is partially true, it overbooks the significant workload and stress the spindle endures. The design of the spindle and the quality of its internal components are vital to its performance and longevity. The spindle is, in many ways, the heart of the CNC machining center.

High-quality components not only determine the spindle's longevity but also affect how it handles speed, torque, and vibration. When researching spindle technologies, you will often find that the bearing system is a key focus. Important considerations when evaluating a CNC machining center's bearing system include the material, the type, the arrangement, and the lufbrication of bearings.

2. Bearing System of the Spindle

In a bearing system (Figure 5.5), the balls roll between the inner and outer steel raceways, and the materials used for the ball bearings affect the spindle's temperature, vibration levels, and the overall longevity. Hybrid ceramic bearings offer distinct advantages over typical steel ball bearings:

(1) Reduced mass.

Ceramic ball bearings have 60% less mass than steel balls. This reduction in mass is particularly significant at high rotational speeds, where centrifugal forces push the balls against the outer race way, potentially leading to increased wear and deterioration of the bearing. However, due to their lower mass, ceramic balls are less susceptible to deformation, allowing for up to 30% higher speed in a given ball bearing size without compromising bearing's longevity.

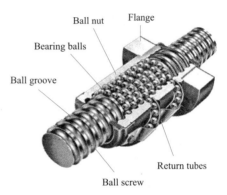

Figure 5.5　A Bearing System

(2) Elimination of cold welding.

Ceramic balls do not react with steel raceways, eliminating the possibility of metal-to-metal contact. This prevents micro scopic or cold welding, a phenomenon where microscopic welding of the ball material to the raceway causes surface wear. As the bearings rotates, these cold welds break, leading to surface roughness, heat generation, and ultimately bearing failure.

(3) Lower operating temperatures.

Due to the nearly perfect roundness of the ceramic balls, hybrid ceramic bearings operate at much lower temperatures than steel ball bearings, which results in longer lubricant life of the bearing.

(4) Reduced vibration levels.

Spindles utilizing hybrid ceramic bearings exhibit higher rigidity and have higher natural frequencies, making them less sensitive to vibration, which results in reduced wear and longer lubricant life of the bearing.

3. Types of Bearings

Various types of bearings are used in the spindle design, with angular contact ball bearings being the most common in high-speed spindle applications (Figure 5.6). These bearings are designed to provide the precision, load-carrying capacity, and the speed necessary for effective metal cutting. Precision balls are fitted into a precision steel race way, offering both axial and radial load-carrying capacity. This makes them ideal for the demands of high-speed CNC machining.

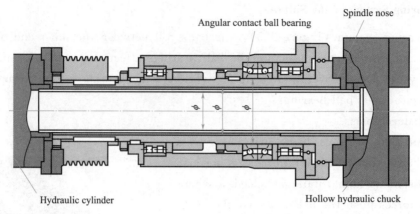

Figure 5.6 Angular Contact Ball Bearing Used in a CNC Lathe

Another type of bearing sometimes used in spindles includes taper roller bearing or cylindrical roller bearing. These bearings offer higher load-carrying capacity and greater stiffness compared to ball bearings, making them suitable for specific spindle speed requirements and applications. Spindle manufacturers often use a combination of these bearing types in different parts of the spindle, depending on the type of load the bearing needs to counteract.

4. Lubrication

Proper lubrication of bearings is essential to ensure their longevity and optimal performance. There are several systems that CNC machine tool manufacturers use to keep the bearings properly lubricated, including oil-mist system, oil-air system, oil-jet system, and pulsed oil-air system.

These systems are particularly necessary when spindle speeds exceed 18000r/min, although they do increase both maintenance and replacement costs. Additionally, these systems require careful monitoring to ensure the correct ratio and the amount of oil and air, or mist.

Permanently lubricated bearings offer a lower-maintenance alternative. With these bearings, lubrication is handled during the assembly of the spindle. Bearings can also be prepacked with grease by the bearing suppliers.

5. Types of Spindles

Spindle technology offers various methods for driving spindles, including belt-driven, gear-driven, inline, and built-in motorized spindles.

When using a belt-driven spindle (Figure 5.7), it is important to ensure that the belt is easy to maintain and accessible to minimize maintenance costs. Additionally, the type of belt can affect the noise level of the machine tool. A belt with a herringbone design is quieter than other belt designs because it disperses trapped air to reduce noise.

Gear-driven spindles add to the cost of the machine tool and can be noisier and require more maintenance than the belt-driven spindle. While gear-driven spindles were once preferred over belt-driven spindles, advances in materials and belt designs have made belt-driven spindles a low maintenance alternative.

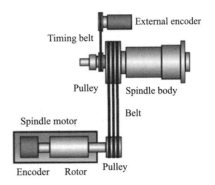

Figure 5.7 A Belt-driven Spindle

The spindle of the inline spindle (sometimes called a direct-driven spindle) is coupled directly to the motor. Inline spindles provide excellent surface finishes and smoother and quieter operation.

Another type of spindle is the built-in motorized spindle, which has the motor integrated into the spindle. These spindles are generally used when higher spindle speeds (in excess of 16000r/min) are required. However, they are more costly compared to belt-driven spindles.

Regardless of the type of spindle, the motor that drives the spindle is critically important. Motors with two sets of windings, known as dual-wound spindle motors, provide more cutting torque and material removal capacities. Single-wound spindle motors are used where lower torques are sufficient, and higher base speeds are not an issue.

6. Enemies of the Spindle

The two major enemies of the spindle are heat and contaminants (namely, chips and coolant invading the bearing system). It is important to identify what design features are included (or available as options) to protect the spindle. Historically, the most common cause of spindle failure has been bearing failure due to contamination from coolant ingress, condensation, or chip damage. To keep the spindle cool and avoid bearing failure, it is crucial to ensure that contaminants stay out.

In most cases, contaminants enter the spindle because the spindle seal has failed. It is important what design measures the machine tool manufacturer has implemented to maintain the seal integrity. An air purge system, which uses a labyrinth seal combined with positive air pressure effectively keeps contaminants out. A dual air purge system, which features two ports (usually upper and lower), is a design that works well to keep contaminants away from the spindle.

Temperature is another factor that leads to spindle problems. Because heat causes steel to expand, manufacturers should explain the measures they have taken to protect the spindle from head growth, which mostly affects the Y and Z axes.

Heat exchangers or chillers (the most common solutions) are used to keep the spindle cool and control both spindle growth and head growth. This type of system extends the longevity of the spindle and reduces head growth, making it useful when running long cycles or high-duty cycles. The selection of the chiller depends on specific applications. For extended high speed applications, you may want to consider a thermal stabilization system, which uses a thermostat and an oil chiller to automatically cool the spindle as needed.

Another factor that impacts spindle performance is the tools being used. Using unbalanced, worn, or excessively long tools can negatively affect the longevity of the spindle.

7. Cooling Considerations

Just as with the spindle, the temperature has a negative impact on the cutting. It is important to check whether the spindle comes with a coolant ring or uses flexible coolant nozzles. When using a coolant ring, it is beneficial to know how many nozzles there are and whether they are adjustable. Obviously, the more nozzles, the better. The ability to adjust the direction of the nozzles is advantageous, as it allows for coverage of a wide range of tool lengths without requiring frequent adjustments.

Coolant through the spindle (CTS) is generally recommended when machining at 12000r/min or faster, particularly if operators are using custom or expensive tools that need to be protected. CTS may also be recommended at lower spindle speeds for certain applications and duty cycles. The price of this feature varies depending on the pressure of the CTS and the spindle design.

8. Replacement Costs

Just like tires on a car, the spindle on a vertical CNC machining center will eventually need to be replaced. When making a purchasing decision for a new CNC machining center, it is important to consider the future costs associated with the spindle replacement. Key considerations include the cost of spindle replacement, the availability of spindle, and the downtime to install.

5.2.2 Guide Rail Design

【拓展视频】

The first frame subsystem design to consider would be a conventional guide rail system, which consists of a linear motion bearing and shaft assembly. This assembly allows unrestricted movement along its lengths. The most logical guide rail design to consider, given the design specifications and size requirements, would be one that could be supported in some way to handle the loads applied to it without much deflections. For instance, the guide rail system shown here features a simple steel shaft railing system that is lightweight. Over the years, there have been vast improvements made in guide rail design to enhance the performances of the guide rail system.

【拓展视频】

Steel shaft railing is both a simple and efficient design for linear motion applications. The steel shaft provides support for loading applications along its length, as well as accommodating the forces generated from linear motion, making this an ideal concept for this particular system.

A guide rail system uses a shaft and support structure to accommodate loading applications along the shaft while managing the forces generated from the linear motion. The shaft and support structure in this particular system can be made from ceramic materials, which offer enhanced properties. These enhanced properties include reduced vibration and minimized deflections of the shaft during loading conditions, both of which contribute to increased shaft longevity.

The guide rail design is highly versatile and has been well-engineered for various loading applications. Each system available, even from different vendors, features guide rail systems that range from ceramic guide rails to case-hardened steel guide rail systems. Most guide rail systems are constructed from case-hardened steel and include some sort of bearing to complement them.

The V-notch guide rail system uses a notch with a V-grooved wheel riding along the railing surface to carry the load and support linear motion. The design of the V-notch guide rail can be made more complex by notching both the top and bottom of the guide rail, allowing it can be used for rails suspended above the ground. This makes the V-notch guide rail system an ideal concept for particular systems (Figure 5.8 and Figure 5.9).

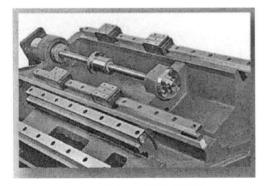

Figure 5.8　Rolling Guide Rail System

Figure 5.9　Sliding Guide Rail System

5.2.3　Rotary Table Design

A rotary table is a precision work positioning device used in the metalworking. It enables the operator to drill or cut workpieces at exact intervals around a fixed axis, which is usually horizontal or vertical. Some rotary tables are equipped with index plates for indexing operations, and others can be fitted with dividing plates that enable precise work positioning at divisions not covered by standard indexing plates. When a rotary fixture is used in this way, it is more accurately called a dividing head (or indexing head).

【拓展视频】

The rotary table shown in Figure 5.10 is a manually operated type.

However, powered rotary tables under the control of CNC machines are now available, providing a fourth axis for CNC milling machines. Rotary tables are constructed with a solid base that can be clamped onto another table or fixture. The actual table is a precision-machined disc to which the workpiece is clamped (T-slots are generally provided for this purpose). This disc can either rotate freely for indexing or be controlled by a worm (handwheel), with the worm wheel portion being an integral part of the rotary table. High-precision rotary tables are driven by backlash-compensating duplex worms.

Figure 5.10　Manual Rotary Table

A rotary table can be used for various applications, including:

(1) Machining spanner flats on a bolt.

(2) Drilling equidistant holes on a circular flange.

(3) Cutting a round workpiece with a protruding tang.

(4) Creating large diameter holes by milling in a circular tool path on small milling machines that lack power to drive large twist drills (over 13mm).

【拓展视频】

(5) Milling helixes.

(6) Cutting complex curves (with proper setup).

(7) Cutting straight lines at any angle.

(8) Cutting arcs.

(9) Using a compound table on top of the rotary table, which allows the user to move the center of rotation anywhere on the part. This enables an arc to be cut at any place on the part.

(10) Cutting circular pieces.

Additionally, when converting into a stepper motor operation and using a CNC milling machine and a tailstock, a rotary table enables the milling of parts that would typically require a lathe.

5.3　Tool System of CNC Machine Tools

An ATC is shown in Figure 5.11. CNC machine tools can be used for:

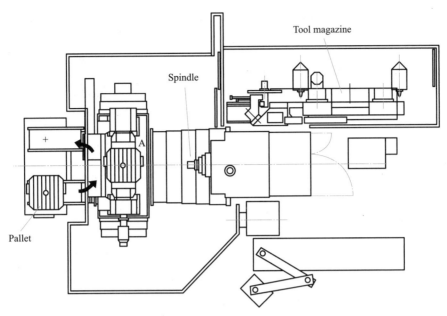

Figure 5.11　An ATC

(1) Parts with complex contours that cannot be manufactured by conventional machine tools.

(2) Small lot production, often even for single (one-off) job production, such as prototyping, tool manufacturing, etc.

(3) Jobs requiring very high accuracy and repeatability.

(4) Jobs requiring many setups and/or where the setups are very expensive.

(5) Parts that are subjected to frequent design changes and consequently require more expensive manufacturing methods.

(6) Situations where the inspection cost is a significant portion of the total manufacturing cost.

5.3.1　Tools

1. Tool Materials

A summary of applications for varicus cutting tool materials is shown in Table 5-1.

【拓展视频】

(1) High speed steel.

(2) Cemented carbides.

Table 5-1　A Summary of Applications for Various Cutting Tool Materials

Tool Material	Work Materials	Remarks
Carbon steels	Low strength, softer materials, non-ferrous alloys, plastics	Low cutting speeds, low strength materials

Continued

Tool Material	Work Materials	Remarks
Low/medium alloy steels	Low strength, softer materials, non-ferrous alloys, plastics	Low cutting speeds, low strength materials
High Speed Steel	All materials of low and medium strength and hardness	Low to medium cutting speeds, low and medium strength materials
Cemented carbides	All materials up to medium strength and hardness	Not suitable for low speed applications
Coated carbides	Cast iron, alloy steels, stainless steels, super alloys	Not for titanium alloys and nonferrous alloys, as the coated grades do not offer additional benefits over uncoated grades
Ceramics	Cast iron, Ni-base super alloys, non-ferrous alloys, plastics	Not for low speed operation or interrupted cutting. Not for machining aluminium alloys and titanium alloys

The following guidelines would be useful for selecting a carbide grade:

(1) Choose a grade with the lowest cobalt content and the finest grain size consistent with adequate strength to eliminate chipping.

(2) Use straight WC grades if cratering, seizure, or galling are not experienced with work materials other than steels.

(3) To reduce cratering and abrasive wear when machining steel, use grades containing TiC.

(4) For heavy cuts in steel, where high temperature and high pressure deform the cutting edge plastically, use a multi-carbide grade containing W-Ti-Ta and/or lower binder content.

(5) Coated carbides.

(6) Ceramics.

The following guidelines would be useful for selecting a ceramic cutting tool:

(1) Use the highest cutting speed recommended, and preferably select square or round inserts with a large tool nose radius.

(2) Use a rigid machine tool with high spindle speeds and a secure clamping angle.

(3) Machine rigid workpieces.

(4) Ensure adequate and uninterrupted power supply.

(5) Use negative rake angles so that less force is applied directly to the ceramic tip.

(6) Keep the overhang of the tool holder to a minimum, no more than 1.5 times of the shank thickness.

(7) Use a large nose radius and side cutting edge angle on the ceramic insert to reduce the tendency of chipping.

(8) Always take a deeper cut with a light feed rather than a light cut with a heavy

feed. Ceramic tips are capable of cutting as deep as half the width of the cutting surface on the insert.

(9) Avoid coolants with aluminium oxide-based ceramics.

(10) Review machining sequence while converting into ceramics, and if possible, introduce chamfer or reduce feed rate at entry.

2. Applications of Tools

Tools are used to remove material from the workpiece by means of shear deformation. Cutting may be accomplished by single-point or multi-point tools. Single-point tools are used in turning, shaping, planning, and similar operations, and they remove materials using one cutting edge. Milling and drilling tools are often multi-point tools. Grinding tools are also multi-point tools, and each grain of abrasive functions as a microscopic single-point cutting edge (though with a high negative rake angle) and shears off tiny chips. The structure of a tool is shown in Figure 5.12.

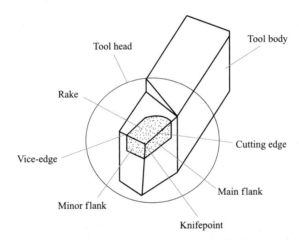

Figure 5.12 The Structure of a Tool

Tools must be made of a material harder than the material to be cut, and the tool must be able to withstand the heat generated during the metal-cutting process. Also, the tool must have a specific geometry, with clearance angles designed so that the cutting edge can contact the workpiece without the rest of the tool dragging on the workpiece surface. The angle of the cutting face, flute width, number of flutes or teeth, and margin size are all important factors. To ensure a long working life, all of these aspects must be optimized, along with the speeds and feeds at which the tool is seperated.

Standard insert shapes (Figure 5.13):

(1) V (35°diamond) —Used for profiling, this is the weakest insert, with 2 edges per side.

(2) D (55°diamond) —Somewhat stronger, used for profiling when the angle allows it, also with 2 edges per side.

(3) T (triangle) —Commonly used for turning because it has 3 edges per side.

(4) C (80°diamond) —A popular insert, because the same holder can be used for both turning and facing, with 2 edges per side.

(5) W (80°triangle) —The newest shape, capable of turning and facing like the C insert, but with 3 edges per side.

(6) S (square) —Very strong, but mostly used for chamfering because it cannot cut a square shoulder, with 4 edges per side.

(7) R (round) —The strongest insert but the least commonly used.

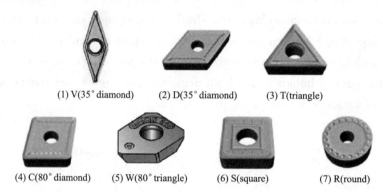

(1) V(35°diamond)　　(2) D(35°diamond)　　(3) T(triangle)

(4) C(80°diamond)　　(5) W(80°triangle)　　(6) S(square)　　(7) R(round)

Figure 5.13　Insert Shapes Manufactured by Sandvik Coromant

Turning (Figure 5.14) is a lathe operation in which the tool removes metal from the outside diameter of a workpiece.

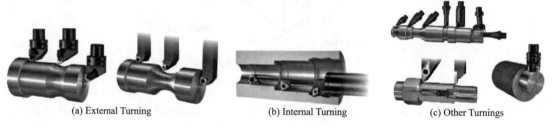

(a) External Turning　　(b) Internal Turning　　(c) Other Turnings

Figure 5.14　Turning

Threading (Figure 5.15) is a process of cutting threads inside a hole so that a cap screw or bolt can be threaded into it. It is used to make threads on nuts.

A vertical milling machine uses a rotating tool to produce flat surfaces. It is a very flexible, light-duty machine tool (Figure 5.16).

Boring is an operation used to enlarge and finish holes accurately, which can be done on a lathe or a milling machine.

Grinding is an operation where cutting is done using abrasive particles. Grinding can remove very small chips in large numbers by the

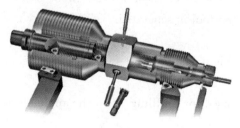

Figure 5.15　Threading

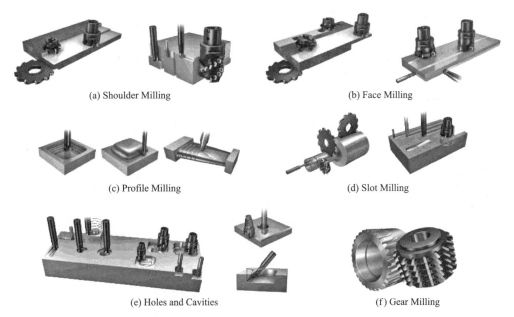

(a) Shoulder Milling (b) Face Milling
(c) Profile Milling (d) Slot Milling
(e) Holes and Cavities (f) Gear Milling

Figure 5.16 Milling

cutting action of many small abrasive grains.

Drilling is an economical way of removing large amounts of metal to create semi-precision round holes or cavities (Figure 5.17).

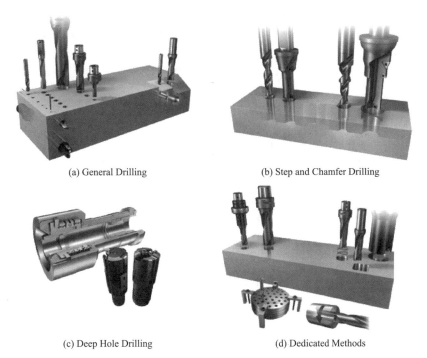

(a) General Drilling (b) Step and Chamfer Drilling
(c) Deep Hole Drilling (d) Dedicated Methods

Figure 5.17 Drilling

Reaming is a sizing operation that removes a small amount of metal from an already drilled hole.

Honing is an internal cutting technique that uses abrasives on a rotating tool to produce extremely accurate holes with a very smooth finish.

3. Tool Holders

Typical applications of tool holders are shown in Figure 5.18.

Figure 5.18 Typical Applications of Tool Holders

Typical tool holders are shown in Figure 5.19, Figure 5.20 and Figure 5.21.

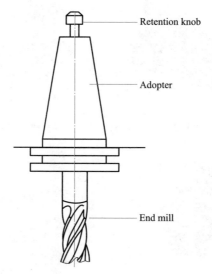

Figure 5.19 Spindle Tool Holder (for an End Mill)

4. Probe System

Probe system (Figure 5.22) is used for a number of applications:

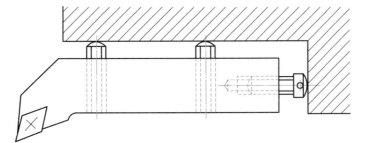

Figure 5.20 Preset Tool Holder Used in CNC Turning Machine

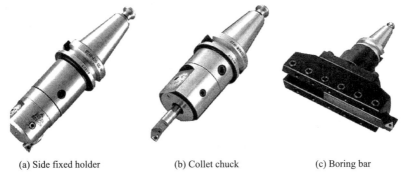

(a) Side fixed holder (b) Collet chuck (c) Boring bar

Figure 5.21 Milling Tool Holders

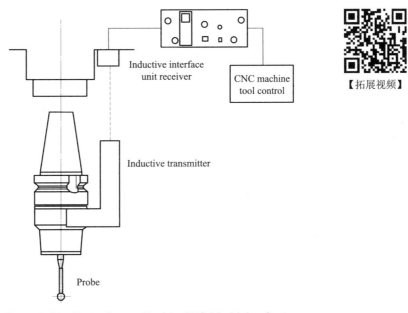

Figure 5.22 Probe System Used in CNC Machining Centers

(1) Datum of the workpiece.
(2) Workpiece dimension measurement.
(3) Tool offset measurement.
(4) Tool breakage monitoring.

(5) Digitizing.

The probe system examples are shown in Figure 5.23.

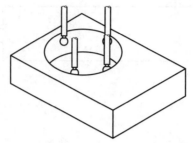

(a) Inspection of a Bore for Diameter and Center Position

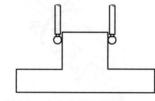

(b) Inspection of a Web Thickness

Figure 5.23　Probe System Examples

5.3.2　Automatic Tool Changer

1. Typical Automatic Tool Changers

CNC machine tools are equipped with controllable ATC. Depending on the type and application area, ATCs can simultaneously hold various quantities of tools and set the tools, called by the CNC programs, into the working position. The most common types are:

(1) Tool turret.

The tool turret (Figure 5.24) is mostly used for lathes, while the tool magazine is used for milling machines. If a new tool is called by the CNC program, the turret rotates until the required tool reaches the working position. Currently, the tool change takes only a few seconds.

Depending on their types and sizes, the turrets of CNC machine tools have 8 to 16 tool positions. In large milling centers, up to 3 turrets can be used simultaneously. If more than 48 tools are needed different types of tool magazines used in such machining centers, allowing a capacity of up to 100 and even more tools. These include longitudinal magazines, ring magazines, plate magazines, chain magazines (Figure 5.25), and cassette magazines.

(2) Tool magazine.

In the tool magazine (Figure 5.26), the tool change takes place in ATC. The change occurs with a double-arm gripping device after a new tool has been called by the CNC program, as follows:

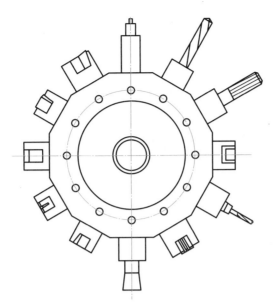

Figure 5.24 Tool Turret

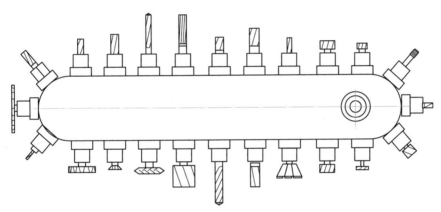

Figure 5.25 A Chain Magazine

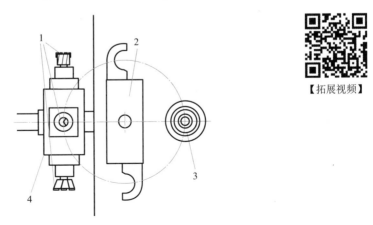

1—milling tools; 2—tool gripper (tool changer); 3—work spindle; 4—tool magazine.

Figure 5.26 A Tool Magazine

(1) Position the desired tool in the magazine into the tool-changing position.

(2) Move the work spindle into the tool-changing position.

(3) Revolve the tool gripping device into the old tool in the spindle and into the new tool in the tool magazine.

(4) Move the tools into the spindle and tool magazine and revolve the tool gripping device.

(5) Place the tools into the spindle sleeve or tool magazine.

(6) Return the tool gripping device to the home position.

The tool change takes 6 to 15s, although ATCs can complete the tool change in merely.

2. Requirement for ATCs

For ATCs to operate, it is necessary to have the following:

(1) A tool magazine where a sufficient number of tools can be stored.

(2) A tool adopter that has a provision for pick-up by the tool change arm.

(3) The ability in the control system to perform the tool-changing function.

(4) A defined tool-changing procedure.

The tool-changing activity requires the following motions:

(1) Stop the spindle at the correct orientation for the tool change arm to pick up the tool from the spindle.

(2) Move the tool change arm to the spindle.

(3) The tool change arm pick up the tool from the spindle.

(4) Index the tool change arm to reach the tool magazine.

(5) Index the tool magazine to the correct position where the tool from the spindle is to be placed.

(6) Place the tool in the tool magazine.

(7) Index the tool magazine to bring the required tool to the tool change position.

(8) The tool change arm pick up the tool from the tool magazine.

(9) Index the tool change arm to reach the spindle.

(10) Place the new tool in the spindle.

(11) Move the tool change arm into its parking position.

Security precautions in CNC machine tools are as follows:

The primary goal of work security is to eliminate accidents and damages to persons, machine tools, and facilities at the work site.

Essentially, the same work security precautions that apply to conventional machine tools also apply to CNC machine tools. These precautions can be classified into three categories:

(1) Eliminate danger.

① Defects on machine tools and all devices necessary for work need to be reported immediately.

② Emergency exits must be kept clear at all times.

③ No sharp objects should be carried in clothing.

④ Watches and rings should be removed before operating machine tools.

(2) Screen and mark risky areas.

① Security precautions and corresponding notifications must not be removed or deactivated.

② Moving and intersecting parts must be properly screened.

(3) Eliminate danger exposure.

① Protective clothing must be worn to guard against potential sparks and flashes.

② Protective glasses or shields must be worn to protect eyes.

③ Damaged electrical cables must not be used.

Exercises

1. Write the type of drills from the box below under the correct picture (Figure 5.27).

| high-helix drill | core drill | gun drill | oil hole drill |
| spade drill | step drill | low-helix drill | straight-fluted drill |

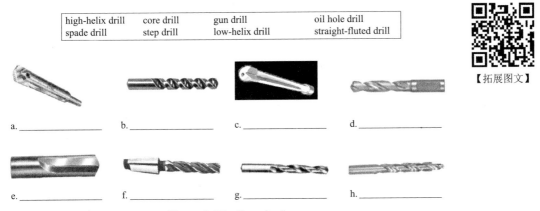

a. _____ b. _____ c. _____ d. _____

e. _____ f. _____ g. _____ h. _____

Figure 5.27 Exercise 1

2. Match each picture with the lathe accessories (Figure 5.28).

a. four-jaw centering chuck   b. steady rest   c. lathe center

d. revolving tailstock center   e. lathe dog   f. boring tools

g. cam-lock spindle nose   h. tool holders   i. three-jaw centering chuck

j. mandrel   k. follower rest   l. tool post

3. Match each picture with the tools and accessories (Figure 5.29).

a. fly tool   b. end mill

c. woodruff key   d. boring head

e. vise   f. shell mill

g. keyseat   h. boring chuck

i. collet   j. slitting saw

k. drill chuck   l. T-slot tool

Figure 5.28 Exercise 2

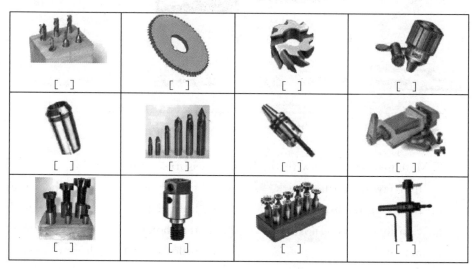

Figure 5.29 Exercise 3

4. Identify the following machining center accessories by writing their names under the pictures (Figure 5.30).

    automatic tool changer   oil-hole drill   two-flute end mills
    single-point boring tool   rose reamer   face milling cutter
    spiral flute tap   precision machine vise

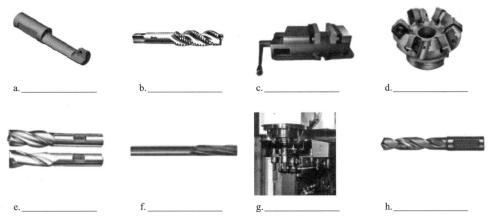

a.＿＿＿＿＿＿＿ b.＿＿＿＿＿＿＿ c.＿＿＿＿＿＿＿ d.＿＿＿＿＿＿＿

e.＿＿＿＿＿＿＿ f.＿＿＿＿＿＿＿ g.＿＿＿＿＿＿＿ h.＿＿＿＿＿＿＿

Figure 5.30　Exercise 4

5. A machining center accessory used for checking part dimensions is called ＿＿＿＿＿＿.

A. A tool set　　　　　　B. A digital readout

C. A trigger probe　　　　D. A dial indicator

6. Narrate the components of a CNC lathe.

7. What are the two major enemies of the spindle?

8. What are the functions of a rotary table?

9. Narrate the main tool materials.

【拓展图文】

Chapter 6
CNC Technology Development

Objectives

- To understand the open CNC system.
- To understand the STEP-NC.
- To understand the advanced applications of CNC technology.
- To understand the intelligent manufacturing technologies.

【拓展视频】

6.1 Open CNC System

After being developed by MIT in the early 1950s, CNC systems have advanced with the emergence and advancement of the microprocessor. With the introduction of automation systems in the 1970s, the functionality of CNC systems has made rapid progress. However, due to the complexity of CNC technology, which requires not only fundamental control functions but also various auxiliary technologies such as machining technology, process planning technology, and manufacturing technology, the market for CNC system has been dominated by a few market leaders in Japan and Germany. The advanced manufacturers evolved CNC systems into closed systems in order to prevent their own technology from leaking out and keeping their market share. However, since the mid-1980s, a new manufacturing paradigm, where computer networks and optimization techniques were applied to manufacturing systems with the progress of computer technology, has emerged along with the requirements for advanced control functions for high-speed and high-accuracy machining.

Closed CNC systems were not adequate for realizing the new manufacturing paradigm. The architecture of closed CNC systems could not meet the users' requirements, and the improvement of CNC system was possible not by MTB (machine tool builders) but by CNC system makers. The limited resources of CNC makers made it impossible to meet the new paradigm.

Therefore, various efforts to develop open CNC systems have been made. As a typical result of these attempts, PC-NC was introduced in the early 1990s. Like IBM PC technology, which appeared in the early 1980s and has progressed by third-party developments based on openness, CNC systems have evolved to PC-NC based on the openness of PC technology. However, now, despite the low price, openness, and many developers of PC-NC, the lack of reliability and openness to application software has made it impossible to implement perfectly open systems (Figure 6.1).

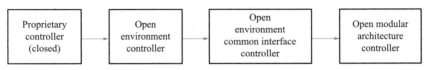

Figure 6.1 Development of Open CNC System

An open system is defined as a system that satisfies the following:

(1) Interoperability. This refers to the ability of the components within the system to cooperate in order to perform the specified tasks. For this ability, standard specifications of the data representation language, behavior model, physical interface, communication mechanism, and interaction mechanism are needed. A bus-based system design is crucial.

(2) Portability. This refers to the ability of a component to be executed on the CNC system with different hardware or software. Portability is very important from a commercial perspective, as it means that a hardware device or software module can be used on various platforms, thereby increasing the efficiency of a platform.

(3) Scalability. This refers to the ability to make extensions to or reductions in the system's functionality possible without incurring significant costs. Adding memory or a board to a PC is a typical example.

(4) Interchangeability. This refers to the ability to replace an existing component with a new one. Instead of replacing the entire system, replacing an existing motion board with a new algorithm is a typical example.

As the definition of an open system, aspects such as modularity, extensibility, reusability, and compatibility can be considered. However, these aspects may seem to belong to the properties listed above. From another perspective, an open system can be defined as a system characterized by flexibility and standardization. Flexibility, though similar in meaning to interoperability and scalability, and standardization, akin to portability and interchangeability, are crucial.

The presented scheme can be used as an economical upgrading method for small or medium industries. From the review of recent researches, the open architecture system is developed due to its flexibility, improved system delivery time, and quality assurance. In this research, an open architecture PC-based numerical control (OAPC-NC) system will be built based on these particular factors. Additionally, the OAPC-NC system should also allow for the easy integration and reuse of hardware and software (Figure 6.2).

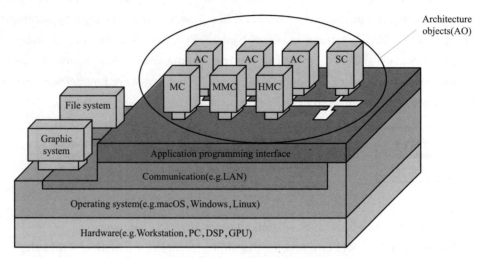

Figure 6.2 OAPC-NC System

Open CNC systems have evolved rapidly. The higher-speed communication options available today have led to many different types of open architecture. Most of these open CNC systems integrate the "openness" of a standard PC with conventional CNC functions. The key advantage of specifying an open CNC system is that it allows the CNC features to remain current with the state of technology and the needs of the process, even as the machine hardware ages. Among the capabilities that can be added to an open CNC system via third-party software, some are more relevant, and some are less relevant where mold machining is concerned. However, across all shops using open CNC systems, some of the most common choices include:

(1) Low-cost network communications.
(2) Ethernet.
(3) Adaptive control.
(4) Interfaces to bar code readers, tool ID readers, and/or pallet ID systems.
(5) Mass program storage and editing.
(6) SPC (statistical process control) data collection.
(7) Documentation control.
(8) CAD/CAM integration and/or shop-floor programming.
(9) Common operator interfaces.

The last item is particularly significant. A growing requirement is for the CNC to be easy to use. An important component of this ease of use is commonality of operation from CNC to another. Typically, operators must be trained separately for separating machine tools because of the CNC interface differences between machine tools and between machine tool builders. Open CNC systems provide new opportunities for working toward a control interface that is common throughout the shop.

Now, machine tool builders can design their own interfaces for CNC operation, and they do not have to be CNC programmers to do so. In addition, open CNC system controls can permit individual log-on so that personnel performing various functions—operator, programmer, maintenance, and so on—see only the screens they need. Eliminating unnecessary screens makes CNC operation even more straightforward.

6.2　STEP-NC

With the rapid advancement of information technology associated with the CNC technology, the manufacturing environment has undergone significant changes over the last decade. However, the low-level standard, G-code & M-code, which have been used for over 70 years as the interface between CAM and CNC, are now considered as an obstacle for global, collaborative, and intelligent manufacturing. A new model of data transfer between CAD/CAM systems and CNC machine tools, known as STEP-NC, is being developed worldwide to replace G-code & M-code.

STEP-NC is expected to encompass the entire scope of E-manufacturing. The new STEP-NC data model has been developed to replace the old standard G-code & M-code for milling, turning, and EDM (electrical discharge machining), and development on implementations is under way. Now that the new data model has been established, the development and implementation of STEP-compliant CAD/CAM/CNC systems based on this new data model are drawing worldwide attention.

As shown in Figure 6.3 (a), G-code contains just spindle speed, instruction, axis movement, tool, feed rate, and miscellaneous function. With this information, it is very difficult for machine tool operators to understand the operational flow, machining conditions, and specification of tools merely by reading a program. Also, it is impossible for the CNC controller to execute autonomous and intelligent control and to cope with emergency cases with this limited information. In contrast, as shown in Figure 6.3 (b), STEP-NC contains the required functional information such as working step, machining operation, machining tool, techndogy, machine function, machining strategy, and machining feature. In other words, STEP-NC includes a richer information set, including both "what-to-make" (geometry) and "how-to-make" (process plan).

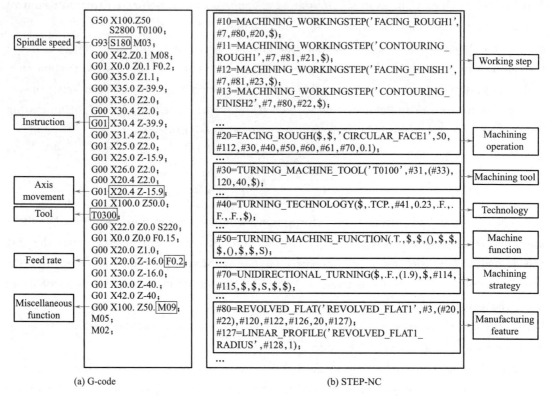

Figure 6.3 Comparison of G-code and STEP-NC

6.3 Advanced Applications of CNC Technology

6.3.1 Components of Flexible Manufacturing System (FMS)

A flexible manufacturing cell (FMC) can be considered as a flexible manufacturing subsystem. The following differences exist between a FMC and a FMS (flexible manufacturing system):

(1) A FMC is not under the direct control of the central computer. Instead, instructions from the central computer are passed to the cell controller.

(2) The cell is limited in the number of part families it can manufacture.

The following elements are normally found in a FMC: a cell controller, a PLC, morethan one machine tools, a materials handling device (a robot or a pallet).

The FMC executes fixed machining operations with parts flowing sequentially between operations.

A FMS consists of two subsystems: a physical subsystem and a control subsystem.

The physical subsystem of a FMS includes the following elements:

(1) Workstations. These consist of CNC machine tools, conventional machine tools,

inspection equipment, loading and unloading operations, and the machining area.

(2) Storage retrieval systems. These act as a buffer during WIP (work-in-process) and include devices such as carousels used to store parts temporarily between workstations or operations.

(3) Material handling systems. These consist of powered vehicles, conveyers, automated guided vehicles (AGVs), and other systems used to transport parts between workstations.

【拓展视频】

The control subsystem of a FMS comprises of the following elements:

(1) Control hardware. This consists of mini and micro computers, PLCs, communication networks, switching devices, and other peripheral devices such as printers and mass memory equipment to enhance the working capability of the FMS.

(2) Control software. This is a set of files and programs used to control the physical subsystems.

The programs, which are created and simulated in the FMS programming system, such as SL FMS (Figure 6.4), can be processed at the machine tool by means of the control module. The operation can be performed alternatively by using a keyboard and/or a mouse. A standard PC keyboard or a control-specific keyboard can be used instead of the numerical control. Modifications of the program must be done in the respective editor of the appropriate programming system. The components and applications of SL FMS is shown in Figure 6.5.

【拓展视频】

Figure 6.4　SL FMS

Basic features of the physical components of the FMS are discussed below:

(1) CNC machine tools. CNC machine tools are considered as the major building blocks of the FMS as they determine the degree of flexibility and capabilities of the system.

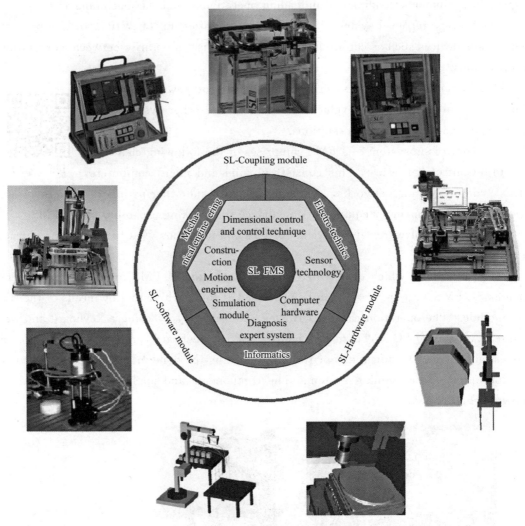

Figure 6.5 The Components and Applications of SL FMS

Some of the features of CNC machine tools are described below:

① The majority of FMSs use horizontal and vertical spindle machining centers. However, machining centers with vertical spindle have lesser flexibility than horizontal spindle machining centers.

② Machining centers have numerical control on movements made in all directions, e.g., spindle movement in X, Y, and Z directions, rotation of tables, tilting of tables, etc., to ensure high flexibility.

③ Machining centers are able of performing a wide variety of operations, such as turning, drilling, contouring, etc. They consist of pallet exchangers interfacing with material handling devices that carry the pallets within and between machining centers as well as automated storage and retrieval systems.

(2) Work holding and tooling considerations. This includes pallets/fixtures, tool

changers, tool identification systems, coolant, and chip removal systems. Key features include:

① Before machining starts on the parts, they are mounted on fixtures. Fixtures must be designed to minimize part handling time. Modular fixing has become an attractive method for quickly fixing a variety of parts.

② The use of automated storage and retrieval system (AS/RS) and material handling systems, such as AGVs, leads to high usage of fixtures.

③ All machining centers are well equipped with tool storage systems called tool magazines. Duplication of the most frequently used tools in the tool magazines is allowed to ensure minimal non-operational time. Moreover, the employment of quick tool changer, tool regrinding, and provision of spares also contribute to efficiency.

(3) Material-handling equipment. Material-handling equipment used in FMSs include robots, conveyers, AGVs, monorails, and other rail-guided vehicles, as well as other specially designed vehicles. Important features include:

① They are integrated with the machining centers and the storage and retrieval systems.

② For prismatic part, material-handling systems are accompanied by modular pallet fixtures. For rotational parts, industrial robots are used to load/unload the turning machine and to move parts between workstations.

③ The material handling system must be capable of being controlled directly by the computer system to direct the materials to various workstations, load/unload workstations, and storage areas.

(4) Inspection equipment. This includes coordinate measuring machines (CMMs) used for offline inspection and programmed to measure dimensions, concentricity, perpendicularity, and flatness of surfaces. A distinguishing feature of this equipment is that it is well integrated with the machining centers.

(5) Other components. Other components include a central coolant and an efficient chip separation system. Their features are:

① The system must be capable of recovering the coolant.

② The combination of parts, fixtures, and pallets must be properly cleaned to remove dirt and chips before operation and inspection.

6.3.2 CNC Technology for Mold Industry

CNC technology is evolving rapidly, which enhances the productivity of machine tools used in the mold industry. Faster CPUs are at the core of many CNC advancements. However, the improvements extend beyond merely quicker processing; the increased speed influences various aspects of CNC technology. Given the significant changes in recent years, it is worthwhile to summarize the state of mold-making CNC technology.

1. Block Processing Time (BPT) and Beyond

As CPU speeds have increased and CNC manufacturers have integrated the speeds into highly sophisticated CNC systems, there have been remarkable enhancements in CNC performance. The faster, more responsive systems do more than just accelerate program block processing. In fact, a CNC system capable of processing program blocks at a very high speed may perform comparably to a system that processes data at a lower speed, because there are other potential bottlenecks downstream that the overall feature content of the CNC system must also address.

Most mold shops today intuitively understand that high-speed machining involves more than just BPT. The analogy of a race car illustrates why this is the case. Should the fastest car win the race? Even a casual observer of racing knows there is more to it than speed alone.

First, the driver's knowledge of the race track is crucial. He must anticipate a sharp curve ahead to decelerate just enough to navigate it safely and efficiently. CNC look-ahead performs a similar role in high-feed-rate mold machining, providing the CNC with advanced knowledge of upcoming sharp turns.

Similarly, how quickly the driver reacts to the actions of other drivers and other unpredictable factors can be compared to the CNC's servo loops (including position loop, velocity loop, and current loop) times.

Consider also the smoothness of the driver's execution as he maneavers around the track. Skillful braking and accelerating have a significant impact on performance. Bell-shaped Acc/Dec in the CNC system provides similar smoothness to machine tool acceleration. Look-ahead is beneficial here as well, because it allows many small Acc/Dec adjustments instead of a sudden change.

The analogy extends further. The power of the engine can be likened to the drives and motors. The weight of the car can be compared to the mass of the moving elements of the machine tool. The strength and rigidity of the car can be equated to the strength and rigidity of the machine tool. Additionally, the CNC's ability to maintain a specified path error can be related to how well the driver keeps the car on the track.

Another way the analogy relates to the current state of CNC technology is this: A car that is not one of the very fastest may not require the most skilled driver. In the past, only high-end CNC systems could maintain high accuracy at high speeds. Today, mid-level and low-end CNC systems are so capable that they may also perform acceptably. The high-end CNC still offers the best available performance. However, for the machine tool, the lower-level CNC will allow the same performance as a top-of-the-line CNC. Previously, the CNC was the limiting factor determining the maximum feed rate in mold machining, but today the limiting factor is the mechanics of the machine tool. A better CNC will not deliver more performance if the machine tool itself is already operating at its performance limit.

2. Features Inherent to the CNC System

Here are some CNC features fundamental to many mold machining processes today:

(1) NURBS interpolation. This feature for interpolating along curves, instead of dividing curves into short, straight line segments, is still gaining popularity. Most CAM packages for die/mold applications now include an option for outputting NURBS-formatted programs. At the same time, more powerful CNC system have allowed CNC manufacturers to add five-axis NURBS capability, as well as NURBS-related features that deliver improved surface finish, smoother motor performance, faster cutting rates, and smaller program sizes.

(2) Finer command unit. Most CNC systems issue motion and positioning commands to machine axes using a command unit of $1\mu m$. Taking advantage of the increase in processing power, some CNC systems today offer a command unit of 1nm. This control increment is 1000 times finer, providing improved accuracy and smoother motor performance, which allows some machine tools to accelerate faster without increasing the shock to the machine tool.

(3) Bell-shaped Acc/Dec. This is also called "jerk control" or "S-curve Acc/Dec". Bell-shaped Acc/Dec allows a machine tool to accelerate faster than linear Acc/Dec. It also results in less position error than various Acc/Dec types, including linear and exponential.

(4) Look-ahead. This is a widely used term, with many performance differences separating the way the feature works on low-end versus high-end controls. In general, look-ahead allows the CNC to preprocess the program to ensure superior Acc/Dec control. The number of look-ahead blocks can range from two to hundreds, depending on the CNC. The number of blocks required depends on the factors such as the minimum program execution time and the Acc/Dec time constant, but a few look-ahead blocks are generally considered the minimum acceptable value.

(5) Digital servo control. Digital servo control technology has improved significantly, and most CNC manufacturers can now offer a digital servo control solution. Advances include faster communications, serial connections between the drive and the CNC, and faster and more numerous digital signal processors. These advances have combined to allow CNCs to control the servo loops more tightly and thus control the machine tool more efficiently. The technology offers numerous benefits in various ways:

① Increased sampling speed of the current loop. With improved current control, the motor heats up less, which not only extends motor's longevity but also reduces heat transfer to the ball screw, thereby enhancing accuracy. Increased sampling speed also enable a higher velocity loop gain, contribute to the overall performance enhancement of the machine tool.

② High-speed serial connection to the servo system. Many modern CNC systems now feature a high-speed serial connection to the servo system, allowing the CNC to receive more detailed information about motor and drive operations. This enhancement has led to improved maintenance features.

③ Serial position feedback. This allows for higher accuracy at high feed rates. As CNC systems have become faster, the rate of position feedback became a limiting in determining how fast a machine tool could move. Conventional feedback, limited by the sample rate of the CNC and the electronics of the external encoder, is now surpassed by serial feedback, which eliminates this bottleneck and allows fine position feedback resolution even at high speeds.

Linear motors technology has seen significant improvements in performance and acceptance in recent years. Advances from companies like GE Fanuc have resulted in machine tool linear motors capable of a maximum force of 15500 N and a maximum acceleration of 30g. Other advancements have led to smaller sizes, lighter weights, and more efficient cooling. These changes enhance the benefits that linear motors offer over rotary motors, such as higher Acc/Dec rates, superior position control, higher stiffness, improved reliability, and inherent dynamic braking.

3. Five-axis Machining

Increasingly applied to complex mold work, five-axis machining (Figure 6.6) technology can reduce the number of setups and/or machine tools required to produce a part, thereby minimizing WIP (work-in-process) inventory and reducing total manufacturing time.

【拓展视频】

【拓展视频】

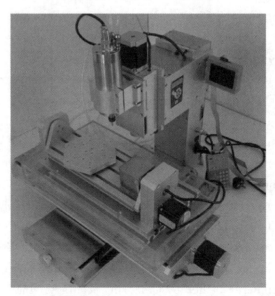

Figure 6.6　Five-Axis Machining

As CNC systems have become more powerful, manufacturers have been able to introduce more five-axis features. These capabilities, once exclusive to high-end controls, are now available in mid-range products. Most of these features simplify five-axis machining for shops with limited experience in this area. Today, accessible CNC technology offers several benefits to the five-axis machining process:

(1) Eliminate the need for qualified tooling.

(2) Allow tool offsets to be set after the program has been posted.

(3) Support "machine anywhere" programming, making posted programs interchangeable from machine to machine.

(4) Improve surface finish.

(5) Support various machine configurations, with the spindle or workpiece pivoting.

One specific feature suited for mold machining is ball-nose end mill compensation. As the part or tool pivots, the CNC dynamically adjusts the tool compensation vector in X, Y and Z axes, maintaining the tool's contact point for a better finish.

Other five-axis CNC functionalities include features related to pivoting the tool, pivoting the part, and allowing the operator to manually move the tool to a new vector.

When rotary axes pivot the tool, the tool length offset, which normally affects only the Z axis, now has components in X, Y and Z axes. In addition, tool diameter offsets, typically affecting only the X and Y axes, also have X, Y and Z components. As the tool feeds in the rotary axes while cutting, all these offsets must be dynamically updated to account for continuous changes in the tool's orientation.

A CNC feature called "tool center point programming" simplifies programming and posting for machine tools achieving rotary-axis motion by pivoting the spindle. This feature allows the programmer to define the path and speed of the tool's center point, while the CNC manages the commands in the rotary and linear axes to ensure accurate tool offsets and eliminate the need for reposting due tool length changes.

Machine tools that achieve rotary motion by pivoting the workpiece use similar functionality, with newer CNC systems compensating for this movement by dynamically adjusting fixture offsets and rotating coordinate axes to match the part's rotary motion.

Additionally, newer CNC systems enhance manual jogging by allowing the axis to be jogged in the direction of the tool vector, facilitating changes in the tool vector without altering the location of the tool tip.

These features collectively make a five-axis machine tool easier to use for "3+2 programming", which is the most common application of five-axis machine tools in moldmaking. However, as new five-axis CNC features continue to evolve and gain acceptance, true five-axis mold machining is likely to become more prevalent.

6.3.3 Virtual Axis Machine Tool

The virtual axis machine tool is known as a hexapod or parallel structured machine tool.

The hexapod, also referred to as a Stewart platform, is a multi-axis machining center capable of full six degrees of freedom (DOF) motion, plus spindle rotation at the tool head. Hexapod machine tools inherit all the advantageous attributes of parallel mechanisms, enabling more potential capabilities for manufacturing. Among these advantages are high structural rigidity, large payload capability, and high speed motions, which are suitable for high-speed and high-accuracy machining.

The designed machine tool can be divided into six basic components:

(1) Main frame.
(2) Moving platform with the main milling spindle.
(3) Telescopic actuators.
(4) System for automatic tool changing.
(5) System for automatic part changing.
(6) Electrical switchbox with control and power components.

Figure 6.7 shows the basic structure of the designed machine tool, including the connection points, places for part changing by the operator, moving platform with the main milling spindle, and electrical switchgear. The main feature is a mechanism with a parallel kinematic structure (PKS) located in the center of the entire machine tool (Figure 6.8).

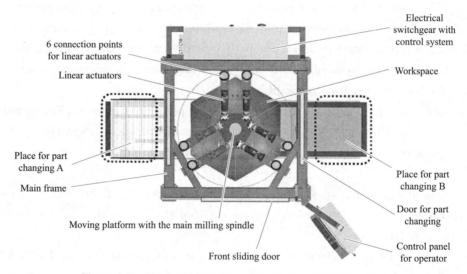

Figure 6.7 The Basic Structure of the Designed Machine Tool

6.3.4 Virtual CNC

The virtual CNC (VCNC) enables the prediction and optimization of a machine tool's dynamic performance at the design stage. By running a program on the VCNC and evaluating the contouring performance, the influence of various design choices such as guideway,

Figure 6.8 Design of a Hexapod Machine Tool with a PKS

drive, encoder, control law, and interpolation algorithm selection can be assessed before the machine tool is actually built.

The VCNC can also be used for tuning servo control and interpolation parameters without taking up production time on the actual machine tool. Once the desired response and contouring accuracy are achieved in the virtual model, the parameters can be directly implemented on the real machine tool with minimal downtime. In process planning, the VCNC can be employed to evaluate the contour errors of different programs and make necessary changes to the feed rate and the tool path to avoid tolerance violations due to servo errors. The simulation accuracy of the VCNC relies on the utilization of realistic mathematical models to describe the dynamic behavior of each component.

Various applications have been developed, which take advantage of the VCNCs accurate simulation capability in predicting and improving the dynamic performance of real CNC machine tools. These applications include:

(1) Prediction of contour errors for programming.
(2) Auto-tuning of servo controllers for feed drives.
(3) Sharp corner tracking using spline interpolation.
(4) Rapid identification of virtual drive models.

The architecture of the VCNC is shown in Figure 6.9, which resembles the real, reconfigurable, and open CNC. The VCNC accepts reference tool path commands generated on CAD/CAM systems in the form of industry-standard cutter location (CL) format. The CL file is interpreted to realize the desired tool motion comprising linear, circular, and spline segments. The axis trajectory commands are generated by imposing the desired feed profile on top of the tool path commands. The feed profiling can be configured to employ piecewise constant, trapezoidal, or cubic acceleration transients.

The axis servo loops are closed by configuring the motion control, feed drive, and feedback modules. The motion controller can be selected from a library of commonly used control laws such as P, PI, PID, P-PI cascade, and lead-lag control, as well as more advanced techniques proposed in the literature, including pole placement, generalized pre-

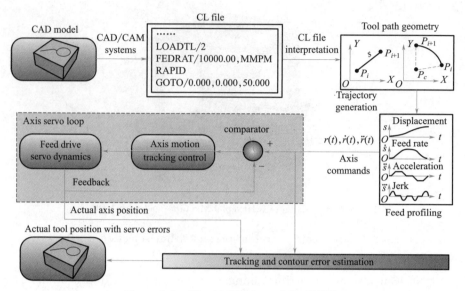

Figure 6.9 The Architecture of the VCNC

dictive control, adaptive sliding mode, feedforward control, and friction compensation.

The feed drive module can be configured to emulate the dynamics of direct or geared drives. Characteristics such as the amplifier, motor, axis inertia, friction, and drive mechanism can be fully defined, including nonlinear effects like quantization, current and voltage saturations, stick-slip friction, and axis backlash. High order drive models with structural resonances either experimentally identified or analytically predicted can also be incorporated. The feedback module can be configured using a combination of linear or angular position, velocity, and acceleration sensors, each with user-defined accuracy and noise characteristics. Once the VCNC is assembled, its performance can be assessed by running various programs and evaluating the servo tracking and contour errors, as well as axis velocity, acceleration, jerk profiles, motor torque, and power histories. Additionally, frequency and time domain analyses can be conducted, which help the user evaluate and improve the stability margins and servo performance of the VCNC axes.

6.3.5 3D Printer

A 3D printer (Figure 6.10) cannot create any object on demand like the "Star Trek" replicators from science fiction. However, a growing array of 3D printers has already begun to revolutionize the business of manufacturing in the real world.

3D printers work by following a computer's digital instructions to "print" an object using materials such as plastic, ceramics, and metal. The printing process involves building up an object one layer at a time until it completes. For instance, some 3D printers extrude a stream of heated, semi-liquid plastic that solidifies as the printer's head moves around, creating the outline of each layer within the object (Figure 6.11).

CNC Technology Development Chapter 6

【拓展图文】

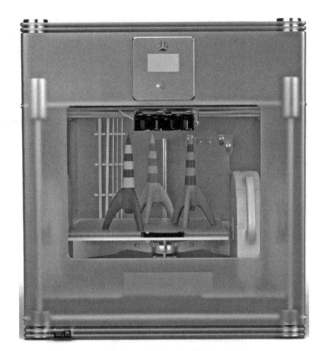

Figure6.10   A 3D Printer

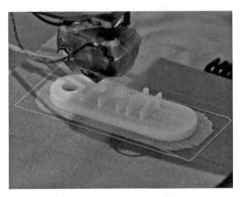

Figure 6.11   3D Printing Sample

The instructions used by 3D printers often take the form of CAD files, which are digital blueprints for making different objects. This means people can design an object on their computers using 3D modeling software, connect the computer to a 3D printer, and watch as the 3D printer builds the object right before their eyes. 3D printing planes are shown in Figure 6.12.

1. History of 3D Printing

Manufacturers have used 3D printing technology, also known as additive manufacturing, to build models and prototypes of products for over 30 years. Charles Hull invented the first commercial 3D printer and offered it for sale through his company, 3D Systems,

Figure 6.12 3D Printing Planes

in 1986. Hull's machine used stereo lithography, a technique that relies on a laser to solidify an ultraviolet-sensitive polymer material wherever the ultraviolet laser touches.

3D printing technology gained wider application until the early 21st century.

A new wave of startups has popularized the idea of 3D printing within the so-called "Maker" movement that emphasizes do-it-yourself (DIY) projects. Many of those companies offer 3D printing services or sell relatively inexpensive 3D printers that can cost just hundreds rather than thousands of dollars.

2. Future of 3D Printing

3D printing probably will not replace many of the traditional assembly-line methods for building standard products. Instead, the technology offers the advantage of making tailored parts on demand, which is more suited for creating specialized parts for military aircrafts. The titanium cabin door of the C919 has used 3D printing to produce more than 28 parts.

The medical industry has also taken advantage of 3D printing's ability to create unique objects that might otherwise be tough to build using traditional methods. For example, researchers built a 3D-printed ear mold that served as the framework for a bioengineered ear with living cells.

The spread of 3D printing technology around the world could also reduce geographical distances for both homeowners and businesses. Online marketplaces already allow individuals to upload 3D-printable designs for objects and sell them anywhere in the world. Rather than paying hefty shipping fees and import taxes, sellers can simply arrange for a sold product to be printed at the nearest 3D printing facility.

6.4 Intelligent Manufacturing Technologies

6.4.1 Artificial Intelligence and Intelligent Manufacturing

1. Artificial Intelligence

Artificial intelligence (AI), sometimes referred to as machine intelligence, is the intelligence demonstrated by machines, in contrast to the natural intelligence displayed by humans and other animals. In computer science, AI research is defined as the study of "intelligent agents": any device that perceives its environment and takes actions that maximize its chance of successfully achieving its goals. The term "artificial intelligence" is applied when a machine mimics cognitive functions that humans associate with other human minds, such as "machine learning" (Figure 6.13) and "problem-solving".

【拓展图文】

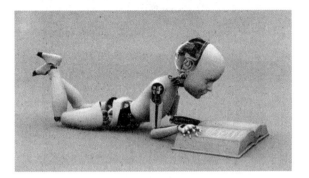

Figure 6.13　Machine Learning

【拓展动图】

The traditional problems (or goals) of AI research include reasoning, knowledge representation, planning, learning, natural language processing, perception, and the ability to move and manipulate objects. General intelligence is among the field's long-term goals. Approaches include statistical methods, computational intelligence, and traditional symbolic AI. Many tools are used in AI, including versions of search and mathematical optimization, artificial neural networks, and methods based on statistics, probability, and economics. The AI field draws upon computer science, mathematics, psychology, linguistics, philosophy, and many other disciplines.

【拓展视频】

Now, the development of AI has stepped into a new stage. With an evolution spanning over 60 years, driven jointly by mobile internet, supercomputing, big data, sensor networks, brain science, new theories, technology, and the strong demands for economic and social development, AI is developing rapidly. It is characterized by deep learning, cross-border fusion, human-machine collaboration, crowd intelligence, and autonomic intelligence. Focus areas include big data-driven knowledge learning, cross-media collaboration processing, human-machine hybrid-augmented intelligence, crowd integration intelligence, and autonomous intelligence systems. The trend towards brain-like intelligence and chip-type hardware and software platforms is becoming more distinct, making a new stage in AI development. Currently, disciplinary development, theoretical modeling, technological innovation, and software & hardware upgrading related to the new generation of AI have advanced, making breakthroughs that promote the rapid advancement of economy, society, and fields tran-

【拓展视频】

【拓展视频】

sitioning from digitization and networking to intelligentization.

The applications of AI include speech cognition, strategic games (such as AlphaGo), autonomous vehicles, intelligent planning, military simulations, artistic performance (Figure 6.14), and more.

Figure 6.14 Artistic Performance

2. Intelligent Manufacturing

Intelligent manufacturing (Figure 6.15) is the development and implementation of AI in the production process. It refers to the use of sophisticated and advanced technologies that can automatically adapt to changing environments and varying production process requirements, with the ability to run manufacturing and assembly lines requiring minimal supervision and reduced manual intervention.

Research areas and focus of intelligent manufacturing include:

(1) Machine vision. 2D/3D defect inspection, intelligent industry robot, 3D vision & cognition, augment reality display for intelligent manufacturing.

【拓展图文】

【拓展视频】

Figure 6.15 Intelligent Manufacturing

(2) Cyber-physical systems (CPSs). Model-based systems engineering for the design, synthesis, and validation of mission-critical CPSs in smart factories under "Industry 4.0".

(3) Sensing. Integrated device modules for manufacturing process, robot, and environmental sensing.

(4) Intelligent energy/power electronics. One-stop product development and service from design, materials, manufacturing processes to performance evaluation and small volume production, etc.

3. Intelligent Manufacturing Technology

Intelligent manufacturing technology (IMT) refers to using computer simulation techniques for intelligent activities in IMS (intelligent manufacturing system). The IMS is based on IMT, incorporating computer application and AI.

【拓展视频】

Compared to traditional manufacturing systems, IMS has several characteristics:

(1) Organizational ability.

In IMS, each composition unit can autonomously build an ultra-flexible optimal structure according to the work duty needs and adapt to the most efficient movement methods. Its flexibility is evident not only its movement but also in its structural design. After completing a task, the structure dismisses itself voluntarily, preparing to build a new structure for the next duty. The voluntary organizational ability is an important symbol of IMS.

(2) Autonomy.

IMS can gather and understand both environmental and internal information, analyze it, and plan its own behavior. A robust knowledge library and a knowledge-based model form the foundation of its autonomy. IMS can monitor and process information according to its environment and working conditions, ultimately self-adjusting to implement the best movement strategies. This autonomy endows the entire manufacturing system with robustness, auto-adaptability, and fault-tolerance.

(3) Self-learning and maintenance.

IMS can continuously improve its knowledge library, using original expert knowledge as a foundation, while eliminating outdated or unsuitable knowledge. This process enhances the efficiency of the knowledge library. At the same time, IMS can perform self-diagnosis, troubleshoot, and repair system failures. Such capabilities enable IMS to optimize and adapt to various complex circumstances.

(4) Intelligent integration of the entire manufacturing system.

While IMS emphasizes the intelligence of each subsystem, it also pays great attention to the intelligent integration of the entire manufacturing system. This is a basic difference between IMS and "isolated intelligent systems" specially applied to manufacturing processes. IMS contains all subsystems and integrates them into a cohesive whole, achieving overall intelligent functionality.

(5) Man-machine integration intelligent system.

IMS is not a purely artificial intelligence system but rather a man-machine integration intelligence system, representing a blend of human and machine intelligence. In this system, human operators play a control role, and human's potential is effectively utilized under the coordination of intelligent machines. It fosters a mutual understanding and collaborative relationship between humans and intelligent machines, allowing them to complement each other at different levels. Consequently, highly skilled and intelligent individuals can perform better roles in IMS, and the integration of machine intelligence with human's wisdom truly enhance the overall system.

(6) Application of Virtual reality.

This technology supports the realization of hypothesized manufacturing processes and facilitates high-level man-machine integration. The union of humans and man-machine represents a new generation of intelligent inter surfaces, transforming available hypothesized methods into reality. This integration is a defining characteristic of IMS. The following will provide a detailed introduction to virtual reality.

6.4.2　Virtual Reality

【拓展图文】

Virtual reality (VR) uses an interactive computer-generated technology that takes place within a simulated environment, incorporating mainly auditory and visual feedback, but also other types of sensory feedback, such as haptic. This immersive environment can resemble the real world or be fantastical, creating experiences that are not possible in ordinary physical reality. Augmented reality (AR) may also be considered a form of VR as they layer virtual information over a live camera feed in a headset or through a smartphone or tablet device, allowing the user to view 3D images.

【拓展视频】

Current VR technology most commonly uses VR headsets or multi-projected environments, sometimes in combination with physical environments or props, to generate realistic images, sounds, and other sensations that simulate a user's physical presence in a virtual or imaginary environment. A person using VR equipment can "look around" the artificial world, move with in it, and interact with virtual features or items. The effect is typically created by VR headsets, which consists of a head-mounted display with a small screen in front of eyes. However, it can also be achieved through specially designed rooms with multiple large screens.

【拓展视频】

VR systems that include the transmission of vibrations and other sensations to the user through a game controller or other devices are known as haptic systems. This tactile information is generally referred to as force feedback in applications such as medical training, video gaming, military training, and driving (Figure 6.16).

Figure 6.16 Virtual Driving

6.4.3 Big Data

Data sets are growing rapidly, in part because they are increasingly gathered by numerous and inexpensive information-sensing devices connected to the internet of things (IOT), such as mobile devices, aerial (remote sensing) technologies, software logs, cameras, microphones, radio-frequency identification (RFID) readers, and wireless sensor networks. The world's technological capita capacity to store information has roughly doubled every 40 months since the 1980s. As of 2012, 2.5 exabytes (2.5×10^{18}) of data are generated every day. By 2025, it is predicted that there will be 491 zettabytes of data every day. One significant challenge for large enterprises is determining who should own big data initiatives that affect the entire organization.

Big data refers to data sets that are so big and complex that traditional data processing application software is inadequate to deal with them. Big data challenges include data capture, data storage, data analysis, search, sharing, transfer, visualization, querying, updating, information privacy, and data source management.

There are a number of concepts associated with big data (Figure 6.17). Initially, there were volume, variety, velocity. Recently, the term "big data" has come to refer more to the use of predictive analytics, user behavior analytics, and other advanced data analytics methods that extract value from data, rather than to the specific size of the data set. The quantities of data now available are indeed large, but this is not the most relevant characteristic of this new data ecosystem. The analysis of large data sets can uncover new correlations, helping to spot business trends, prevent diseases, combat crime, and more. Scientists, business executives, medical practitioners, advertisers, and governments alike face challenges with large data sets in areas such as internet search, fintech, urban informatics, and business informatics. Scientists encounter limitations in e-science fields, including meteorology, genomics, connectomics, complex physics simulations, biology, and environmental research.

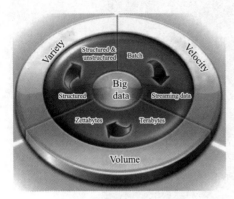

Figure 6.17 Big Data

The rise of cloud computing has paved the way for the cost-effective storage space and computing resources needed for big data. Essentially, cloud computing is a form of outsourcing computer programs. Using cloud computing, users can access software and applications from anywhere, as the computer programs are hosted by an outside party and reside "in the cloud". This means that users do not have to worry about aspects like storage and processing power; they can simply enjoy the end results.

Exercises

1. Narrate the open CNC system.
2. What are the components of FMS? What is the function of each component?
3. What are the types of FMS? What is the feature of each type?
4. What is the concept of 3D printing?

第1章 数控技术概论

教学目的及要求

了解数控技术的发展历程；

了解数控技术与计算机数控技术；

了解数控机床的分类；

了解数控机床与计算机数控机床的特性。

1.1 数控技术的发展历程

数控技术是近现代发展起来的一种自动控制技术，采用数字程序控制来实现机床运动的自动控制。它以数字形式预先记录加工程序和各坐标轴运动方向、速度等信息，同时通过数控装置自动控制机床运动。它还具有自动换刀、自动测量、润滑和冷却等功能。

数控技术起源于1947年，美国帕森斯公司受美国空军委托，研制直升机螺旋桨叶片轮廓样板的加工设备。由于样板的制造难度较大，难以按时完成，因此该公司提出了利用计算机控制机床的设想。

1949年，美国空军意识到飞机和导弹的部件变得越来越复杂，设计不断返工，图纸修改不断。为加快生产，美国空军与承包商帕森斯公司和分承包商麻省理工学院伺服机构实验室一起开始数控机床的研究。1952年，其试制成功第一台由大型立式仿形铣床改装而成的三坐标数控铣床，不久即开始投入生产。

如今，数控机床的发展完全依赖于数控系统。自1952年美国试制成功第一台数控铣床以来，数控系统经历了两阶段和六代的发展。

1. NC 阶段（1952—1970）

由于早期数控装置计算慢，科学计算和数据处理能力低，因此人们不得不采用数字逻辑电路"搭"成一台机床专用计算机作为数控系统。该阶段的数控系统发展分为三代。

第一代数控系统（1952—1959）：数控装置采用电子管元件。

第二代数控系统（1959—1965）：数控装置采用晶体管元件。

第三代数控系统（1965—1970）：数控装置采用小型或中型集成电路。

2. CNC 阶段（1970 年开始）

20 世纪 70 年代通用小型计算机大规模生产，其计算速度与五六十年代相比有很大提升。通用小型计算机与机床专用计算机相比，成本低、可靠性好，很快成为数控装置的核心器件，数控机床进入计算机数控（CNC）阶段。随着计算机软、硬件技术的发展，该阶段的数控系统发展也分为三代。

第四代数控系统（1970—1974）：小型计算机控制系统和大规模集成电路广泛使用。

第五代数控系统（1974—1990）：数控装置采用微型计算机。

第六代数控系统（1990 年至今）：PC 性能大幅提升，已经可以满足数控系统核心部件的性能要求，数控系统进入基于 PC 的开发时代。智能数控装置及智能机床正在发展中，它融合了智能传感、物联网和数字孪生等技术。

1.2　数控技术与计算机数控技术

1.2.1　数控技术

1. 数控

数控是一种可编程自动化的形式，机床或其他设备的机械运动由字母、数字和数据的编码程序控制。这些代码表示主轴头和工件之间的相对位置及操作机床所需的其他指令。主轴头指切削刀具或其他加工装置，工件指被加工的对象。当前作业完成后，可以更改加工程序，以处理新的加工过程。中、小批零件的数控加工只需修改零件的加工程序，就可以加工不同零件。程序更改比硬件设备更改要简单得多。

2. 数控机床的组成

数控系统可以完成常规机床的操作人员需要完成的大多数任务。数控系统必须知道何时以何种顺序发出指令来更换刀具，机床应以何种速度和进给来完成加工，保证零件加工到所要求的尺寸。数控系统通过控制程序（数控程序或零件程序）的数字输入信息以执行各种控制。

数控机床由信息载体、机床控制单元（数控装置）、伺服系统、机床本体和传感器反馈系统组成，如图 1.1 所示。

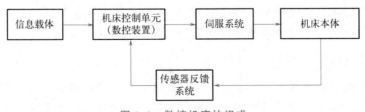

图 1.1　数控机床的组成

数控机床的工作原理如图 1.2 所示。在数控机床上加工零件时，首先根据零件图和数控手册按指定的控制格式开始编写数控程序，数控程序包括所有的加工信息和控制信息；然后将数控程序输入机床控制系统，最常用的方法是通过机床操作面板将数控程序输入数控装置中，计算机将每条数控指令转化为伺服系统对应的控制信号；最后伺服系统控制数控机床按所要求的轨迹运动，完成零件的加工。

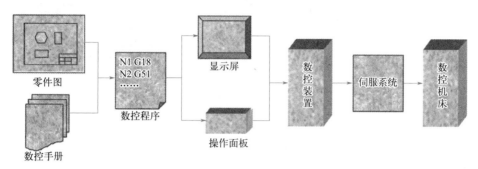

图 1.2　数控机床的工作原理

1.2.2　计算机数控技术

1. 计算机数控系统

20 世纪 70 年代，数控装置进入了小型计算机化时代。随着电子技术和计算机技术的发展，现在 CNC 系统采用微处理器系统和可编程逻辑控制器，它们以并行和联动的方式工作。

CNC 机床在 NC 机床的基础上加上了机载计算机。数控机床的机床控制单元（MCU）通常是"硬"接线的，这意味着所有机床功能都由内置在控制器中的物理电子元件控制。而机载计算机是"软"接线的，这意味着机床功能在制造时被编码到计算机中，并且在关闭 CNC 机床时不至于被删除。保存这些信息的计算机存储器称为只读存储器（ROM）。

MCU 通常有一个字母数字键盘，用于数控程序的输入。这些数控程序存储在计算机的随机存储器（RAM）中，数控程序可以回放、编辑和处理。但是，当 CNC 机床关机时，RAM 中的所有数控程序都会丢失。这些数控程序可以保存在辅助存储设备（如穿孔带、磁带或磁盘）中。最新的 MCU 有图形显示屏，不仅可以显示数控程序，还可以显示生成的刀具路径和数控程序中的错误。

2. CNC 机床的组成

CNC 机床由下列部分组成（图 1.3）。

（1）CNC 装置。

CNC 装置是 CNC 系统的核心，其功能是处理输入的零件加工程序或操作指令，然后将其输出到相应的执行机构中，完成零件加工程序和操作所需的工作。它主要由专用计算机或通用计算机、位置控制单元、人机界面、通信接口界面、扩展功能模块和相应的控制软件组成。显示单元作为机床和操作者之间的交互设备，当 CNC 机床工作时，显示单元

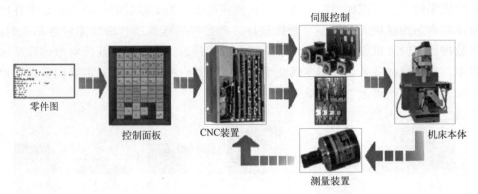

图 1.3 CNC 机床的组成

显示当前机床状态，如机床导轨的位置、主轴转速、进给速度、加工程序等。

在高档的 CNC 机床中，显示单元可以显示刀具轨迹的图形仿真过程，从而在实际加工前验证加工程序的正确性。有关 CNC 机床维护和安装时的许多重要信息（如机床参数、程序控制的逻辑图、报警信息和诊断数据等）也可以显示在显示屏上。

(2) 伺服系统驱动装置和测量装置。

伺服系统驱动装置包括主轴伺服驱动装置、主轴电动机、进给伺服驱动装置和进给电动机等。伺服系统测量装置包括位置控制装置和速度控制装置，是完成主轴控制、进给速度控制和位置控制必不可少的设备。主轴伺服驱动装置主要影响 CNC 机床的切削运动，控制参数是速度；进给伺服驱动装置完成轮廓控制，控制参数是位置和速度。伺服系统的性能主要取决于位置控制装置和速度控制装置的灵敏度和准确性。

(3) 控制面板。

通过控制面板，操作者可以实现和 CNC 机床的交互，完成机床操作、编程和调试，可以方便地设置和修改机床参数，可以了解、查询机床的运动状态。控制面板是 CNC 机床的输入输出设备。

(4) 控制介质和程序输入输出设备。

控制介质是记录零件加工程序的介质，也是人机之间信息交换的载体。程序输入输出设备的作用是实现 CNC 系统与外部设备的信息交换，其将控制介质上记录的零件加工程序输入 CNC 系统，并用输出装置将调试好的零件加工程序存储或记录在适当的介质上。CNC 机床的控制介质和程序输入输出设备是磁盘和磁盘驱动器等。

(5) 机床本体。

机床本体是 CNC 系统加工过程的执行机构。它由主运动部件、进给运动部件、轴承架及专用机构、工作台自动更换系统、自动换刀装置及附属装置等组成。

3. 数控程序输入方法

输入控制介质包含数控程序（指令程序），它指挥 CNC 机床逐步按程序动作，每个指令包括刀具相对于夹有工件工作台的位置和其他附加信息，如主轴转速、进给量、刀具选用和其他信息。程序记录在相应的输入控制介质上，以传送到 MCU。过去长期使用的该介质是穿孔带，它采用 MCU 能够识别的格式排列。如今，穿孔带已基本被新的存储技术

（如磁带、磁盘和计算机网络传输等）所取代。

（1）软盘。

软盘是一种用于 CNC 数据输入的小型磁存储设备。从数据传输速度、可靠性、存储容量、数据处理和读写能力等方面来看，它一直是自 20 世纪 70 年代最常见的存储介质。此外，只要操作者有合适的程序来读取软盘中的数据，就可以在任何时候对其进行编辑，非常方便。然而，软盘的缺点是其不适用于长期使用（软盘容易降解），并且不适合在磁场强及灰尘大的车间和作业场合使用（划痕对软盘的使用是致命的）。

（2）USB 闪存驱动器。

USB 闪存驱动器（俗称 U 盘，图 1.4）是一种可移动和可重写的便携式硬盘驱动器。与软盘相比，它体积小、容量大，常用于没有足够内存或存储缓冲区来存放大型数控程序的机床。

（3）串行通信端口。

计算机和 CNC 机床之间的数据传输通常通过串行通信端口完成。串行通信端口有国际标准，以便有序交换信息。计算机和 CNC 机床之间最常见的端口是 EIA（电子工业协会）的 RS-232 端口。大多数 PC 和 CNC 机床都内置 RS-232 端口，并使用标准的 RS-232

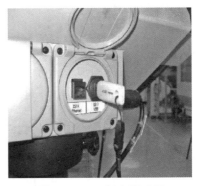

图 1.4　USB 闪存驱动器

电缆将 CNC 机床连接到计算机上，使数据可靠传输。在计算机上运行通信程序，实现 CNC 机床控制器与计算机通信，将数控程序下载到机床存储器中或上传到计算机备份。

分布式数字控制（DNC，图 1.5）是指将一组 CNC 机床连接到一个通用存储器的系统，用于数控程序或加工程序存储，并按需向 CNC 机床分发数据。数控加工程序一次下载一段到控制器中。下载的程序一旦执行，该段程序将被删除，以便为其他程序留出空间。这种方法通常用于没有足够内存或存储缓冲区来存放大型数控加工程序的机床。

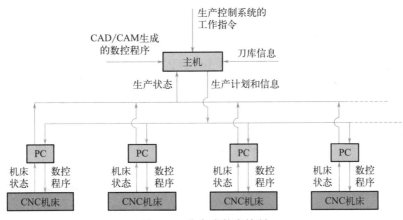

图 1.5　分布式数字控制

DNC 可以看作一个用于在生产管理主机和 CNC 机床之间分配数据的分层系统。通过主机与多台 CNC 机床相连，或 PC 直接与 CNC 机床相连，执行加工程序下载。上位机的通信程序可以利用双向数据传输功能进行生产制造信息（如生产计划、零件工艺和机床使用率等）的通信。

（4）以太网通信。

随着计算机技术的发展和计算机成本的大幅降低，通过以太网通信电缆在计算机和CNC机床之间传输加工程序变得越来越实用且经济可行。这种传输方式为加工程序的传输和存储提供了一种更有效、更可靠的手段。现在，许多公司都建立了局域网（LAN），越来越多的CNC机床为局域网通信安装了以太网卡。

（5）交互式程序设计。

我们可以通过键盘将数控程序输入控制器中。控制器内置的智能软件使操作者能够通过交互逐步输入所需的程序，这是一种非常有效的方法，可以为2.5轴及加工相对简单的零件编制数控程序。

4. CNC机床的加工流程

CNC机床的加工流程（图1.6）如下。

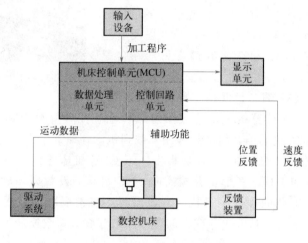

图1.6　CNC机床的加工流程

（1）使用G代码和M代码编写加工程序。加工程序描述了机床为加工零件必须执行的操作顺序。数控编程可以离线编写，即可以脱机（机床）编程，包括手工编程和CAD/CAM软件自动编程。

（2）将加工程序下载到MCU中。在此阶段，仍可以使用MCU的键盘、输入设备编辑或进行程序仿真。

（3）MCU处理加工程序，并向数控装置发送信号，引导机床完成加工零件所需的操作顺序。

无论是手动操作CNC机床还是全自动操作，都可以按加工程序来高精度完成加工循环。

1.3　数控机床的分类

数控机床可以按下列方法进行分类。

(1) 按加工工艺类型分类。
(2) 按机床运动轨迹分类。
(3) 按伺服系统类型分类。

1.3.1 按加工工艺类型分类

数控机床广泛用于金属切削加工。数控加工适用于加工以下类型的零件。
(1) 形状复杂的零件。
(2) 公差精度高或重复性好的零件。
(3) 在普通机床上加工需要昂贵工装夹具的零件。
(4) 存在工程变更的零件，如处于原型试制阶段的零件。
(5) 加工精度极易受操作者个人因素干扰的零件。
(6) 加工任务很急的零件。
(7) 小批生产或短线制造的零件。

常见的数控机床和设备如下。
(1) 数控折弯机（图1.7）。
(2) 钻削加工中心。
(3) 车削加工中心。
(4) 铣削加工中心。
(5) 数控转塔冲床。
(6) 电火花加工机床（图1.8）。
(7) 数控磨床。
(8) 火焰切割机。
(9) 激光切割机（图1.9）。
(10) 超高压水切割机（图1.10）。
(11) 电解加工机床。
(12) 激光测量机（图1.11）。
(13) 三坐标测量机。
(14) 工业机器人。

图1.7 数控折弯机

图1.8 电火花加工机床

图1.9 激光切割机

图 1.10　超高压水切割机

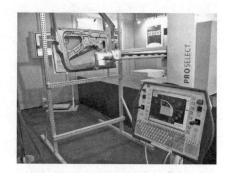
图 1.11　激光测量机

1.3.2　按机床运动轨迹分类

有些 CNC 机床在工件的离散位置执行加工（如钻孔、冲压和点焊），有些 CNC 机床则通过工作台移动进行加工（如车削、铣削和连续电弧焊）。刀具在移动过程中，需要遵循直线、圆弧或其他曲线路径，这些不同类型的移动都是由运动控制系统完成的。

CNC 机床的运动控制系统可以分为点到点控制系统、直线切削控制系统和轮廓控制系统。

1. 点到点控制系统

点到点控制系统（point-to-point control system）又称点位控制系统，刀具从起点向终点移动时，无论其中间的移动轨迹如何，都要求刀具最后能准确到达终点（移动到加工位置），完成指定的加工（如钻孔、冲孔）。加工程序包含一系列的坐标定位，通过点到点控制系统完成各点位的加工，如图 1.12 所示。

数控机床点到点控制系统可以控制不同的坐标轴运动，通常先移动 X 轴，再沿 Y 轴和 Z 轴移动。点到点控制系统的运动轨迹往往是 Z 字形的，直到到达程序指定的终点位置。

2. 直线切削控制系统

多数 CNC 机床的直线切削控制系统（straight-cut control system）都有一个更为复杂的 MCU。在直线切削控制系统中，位置指令是同时控制的，以便完成两轴的矢量运动，如图 1.13 所示。然而，该矢量运动控制只能逐个进行脉冲输出。直线切削控制系统可沿与坐标轴成 45°的斜线进行切削。

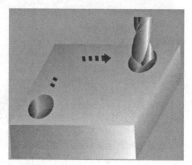
图 1.12　点到点控制系统

图 1.13　直线切削控制系统

3. 轮廓控制系统

轮廓控制系统（contouring control system）可以实现 CNC 机床在给定进给量下的任意轨迹控制。轮廓控制系统又称连续控制系统，这类控制系统可同时对两个或两个以上（若有第四轴和第五轴，常指偏转角运动）的运动坐标位移及速度进行连续控制，起点与终点间的轨迹是通过插补完成的，如图 1.14 所示。

当轮廓控制系统仅用于控制刀具或工作台以所要求的速度沿平行于某一坐标轴方向进行直线切削时，这种控制方式称为直线控制。当对两个或两个以上的运动坐标的位移及速度进行连续控制时，这种控制方式称为轮廓控制。CNC 机床轮廓控制系统能实现直线插补和圆弧插补。

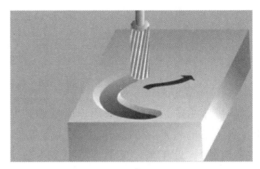

图 1.14　轮廓控制系统

4. 插补

轮廓控制中一个重要概念是插补（interpolation）。轮廓控制系统需要生成的路径通常由圆弧和其他光滑的非线性形状组成。有的形状可以用相对简单的几何公式进行数学定义，而有的形状不能用数学定义，除非用逼近法。在任何情况下，使用 CNC 机床生成这些形状的一个基本问题是它们是连续的，而 CNC 是数字控制，要沿圆形路径加工，必须将圆形分割成许多与圆形路径近似的直线段。指令控制刀具连续加工每个直线段，使加工表面逼近所需形状。名义（理想）表面和实际（加工）表面之间的最大误差由单个线段的长度决定。

如果程序员要为每个线段指定端点，那么编程任务将非常巨大，错误在所难免。此外，由于点数较多，零件加工程序将非常长。为了减轻编程工作量，程序员开发了数控插补程序，通过计算中间点，由 CNC 机床生成特定的数学定义或近似路径。如下插补方法可用于处理生成平滑连续路径时遇到的各种问题。

(1) 直线插补。

(2) 圆弧插补。

(3) 螺旋插补。

(4) 抛物线插补。

(5) 三次样条插补。

每个插补过程都需要程序员使用相对较少的输入参数为线性或曲线路径生成程序指令。MCU 中的插补模块执行数值计算，并引导刀具沿路径运动。插补一般由数控系统软件完成。现代数控系统一般都有直线插补和圆弧插补，螺旋插补也较常见；而抛物线插补和三次样条插补一般很少见，多用于加工复杂表面的轮廓。

1.3.3　按伺服系统类型分类

当传感器检测到的实际速度和位置反馈到 CNC 机床控制电路时，伺服电动机进行实时控制，使速度误差或位置误差最小化。反馈控制系统由 CNC 机床各坐标轴的三个独立控制回路组成：最外圈控制回路为位置控制回路，中间控制回路为速度控制回路，最内圈控制回路为强电控制回路，如图 1.15 所示。一般来说，位置控制回路位于数控装置中，其他回路位于伺服驱动装置中。对于控制回路的位置没有绝对的标准，可以根据设计者的要求改变。

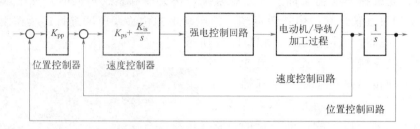

图 1.15　数控机床的控制回路

在机床主轴系统中，采用速度反馈控制，使机床保持给定的转速。反馈信号通常有两种：测速发电机产生的感应电压（模拟信号），作为反馈信号；光电编码器产生的脉冲（数字信号）。近年来，反馈信号更多为数字信号而不是模拟信号。

1. 开环伺服系统

开环伺服系统（open-loop servo system）没有系统实时数据的反馈，因此在系统受到干扰时，不能立即响应并采取纠正措施。开环伺服系统通常只适用于输出几乎是恒定、可预测的情况。开环伺服系统不适用于控制 CNC 机床，因为 CNC 机床的切削力和负载是实时变化的。唯一的例外是线切割机，因为线切割机几乎不存在切削力变化的问题。

2. 半闭环伺服系统

半闭环伺服系统（semi-closed loop servo system）采用测量反馈装置，以确保工作台移动到所需的位置。对伺服电动机输出系统来说，它是间接反馈。虽然这种控制方法在数控系统中很常用，但其不如直接反馈精确。半闭环伺服系统比较指令位置信号与伺服电动机的驱动信号。

在实际操作中，半闭环伺服系统控制工作台移动到笛卡儿坐标系的指定位置。大多数位置控制系统至少有两个轴，每个轴都有一个控制系统，本书只说明其中一个轴的控制系统。与丝杠相连的伺服电动机是每个轴的执行机构，坐标值的指令信号从单片机发送到驱动丝杠的电动机，丝杠的旋转运动被转换为工作台的直线运动。安装在伺服电动机轴或丝杠上的位置传感器用于测量实际位置，当工作台靠近设定的坐标值时，实际位置和输入值之间的差值会逐渐减小。半闭环伺服系统调整方便，稳定性好；但由于负载变化，该系统往往无法检测到齿隙或丝杠失动。

半闭环伺服系统是最常用的控制系统。在这种类型控制中，位置传感器连接到伺服电动机的轴上，检测其旋转角度。轴的相互位置精度对滚珠丝杠的精度有很大的影响，因此，其促进了高精度滚珠丝杠副的研发和广泛应用。滚珠丝杠的高精度弥补了间接测量的精度误差。

如果有必要，则可在数控系统中进行螺距误差补偿和齿隙补偿，以提高定位精度。螺距误差补偿方法是在给定的螺距下，数控系统调用间隙补偿程序，自动将间隙补偿值加到由插补程序给出的位置增量命令中，以消除累积的位置误差。齿隙补偿方法是当运动方向改变时，向伺服驱动系统发送与齿隙量相对应的附加脉冲。近年来，大螺距高导程滚珠丝杠副在高速加工中的应用日益增多。

3. 闭环伺服系统

在闭环伺服系统（closed-loop servo system，图1.16）中，测量反馈装置直接测量工作台的位移输出，能够第一时间反映被测目标的变化。因此，该系统精度较高。其比开环伺服系统要求也更高，适用于输出频繁变化的场合。现在，大多数CNC机床都采用闭环伺服系统。

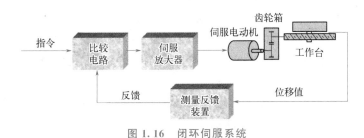

图1.16 闭环伺服系统

半闭环伺服系统和闭环伺服系统通常使用可变速交流电动机或直流电动机，即采用伺服系统控制、驱动各坐标轴。

1.4 数控机床与计算机数控机床的特性

1.4.1 数控机床的优缺点

1. NC机床

NC机床的优点主要体现在其应用上，主要有以下几点。

（1）更高的准确性和可重复性。与普通机床相比，NC机床减少或消除了操作者的技能差异、疲劳和其他个体因素造成的不确定性。NC机床加工的零件尺寸更接近公称尺寸，并且批量生产中零件尺寸变化较小。

（2）可加工常规方法难以加工的复杂表面。数控技术使加工几何形状复杂的零件成为可能，超出了普通加工方法的适用范围，使产品设计有更多的选项。单个的零件可以设计更多的功能特征，以减少产品中的零件总数和相关的装配成本；可以精确制造用数学描述的表面，设计师的挪移空间扩大，可以拓展更多的想象力，设计新的零件和几何模型。

（3）减少了非作业时间。NC机床没有进行金属切削加工工艺的优化，但它增加了作业时间在工时中的占比。NC机床在进行金属切削加工时，减少了工件的加工时间，并且部分NC机床实现了自动换刀，这一优势节约了劳动力成本，并减少了加工零件时所需的时间。

（4）降低废品率。由于更高的加工精度和程序可重复性，以及生产过程中人为误差的减少，NC机床能够制造更多满足公差要求的零件。因此，在生产安排中，可以考虑降低废品率指标，减少每批生产中的返工量，从而节省生产时间。

（5）降低质检工作量。当使用NC机床加工时，检验工作很少，零件的加工程序相同，生产的零件实际上是相同的。一旦程序得到首件验证，就无须采用普通加工方法所需的高频次抽样检查。除非刀具磨损和设备故障，NC机床在加工的每个循环中都会精确加工出相同的零件。

（6）可以更方便地适应设计变更。零件设计的更改往往需要重新设计复杂的工装夹具；而采用 NC 机床加工，只需修改加工程序就可以适应设计更改。

（7）工装夹具简单。加工形状复杂的零件不需要复杂的工装夹具，因为 NC 机床可以实现刀具的精确定位，刀具位置不需要工装夹具来保证。

（8）缩短制造周期。使用 NC 机床可以更快地设置生产计划，而且每个零件所需的搬运次数更少，这样大大缩短了从获得订单到完成订单的时间，即周期。

（9）减少了零件的库存。由于减少了零件的搬运次数，工装夹具的周转、使用简单快速。NC 机床适用于小批量生产零件，对批量的要求比传统加工要宽松，因此，零件的库存得以减少。

（10）厂房占地面积少。与传统机床相比，执行相同加工量所需的 NC 机床更少，减少零件的库存也有助于减少厂房占地面积。

（11）操作者的技能要求降低。操作 NC 机床的个人技能要求通常低于操作传统机床的个人技能要求。NC 机床的操作通常只包括装卸零件和定期更换刀具。加工循环由数控程序完成。某些 NC 机床的换刀也可以通过编程实现。在传统机床中完成相应的操作需要更多的人工参与，操作者需要接受更多的培训以拥有高水平技能。

但是，NC 机床有下列缺点。

（1）更大的资金投入。NC 机床具有比传统机床更高的初期资金投入，原因如下。

① NC 机床包括数控装置和电子硬件。

② 数控装置和软件开发成本包含在设备成本中。

③ NC 机床一般采用更可靠的机械零件。

④ NC 机床通常具有传统机床不具备的附加功能，如自动换刀装置（ATC）和自动上下料机构。

（2）维护工作量大。一般来说，NC 机床需要比传统机床更高的维护水平，这意味着更高的维护和维修成本，这主要是因为现代数控系统中包含计算机和其他电子设备。维护人员必须接受相关设备维护和维修培训。

（3）零件数控编程。NC 机床必须编程。客观地讲，无论是否在 NC 机床上生产，任何零件加工都必须考虑工艺规划。然而，数控编程是批量生产中一个特殊的准备步骤，在传统车间不存在编程问题。

（4）为提高 NC 机床的利用率，使 NC 机床的经济效益最大化，通常需要采用倒班制，这意味着管理人员和其他员工需要为企业的正常运行增加一到两个额外的班次来工作。

2．CNC 机床

相对 NC 机床，CNC 机床的优点有如下几点。

（1）减少了机床功能扩展所需的硬件投入。新的功能以软件方式集成到 MCU 中，通过编程实现。

（2）加工程序可以直接被 CNC 机床写入、存储和执行。

（3）CNC 机床加工程序的任何一部分都可以回放和编辑，而且回放时可以显示刀具运动轨迹。

(4) MCU 可以存储大量不同类型的加工程序。

(5) 多台 CNC 机床可以与主机联网。加工程序可以通过主机下载到联网的 CNC 机床中，即 DNC。

(6) 多个 DNC 系统可以向上联网，组成大型 DNC 系统。

(7) 数控程序能够通过 USB 闪存驱动器、软盘和磁盘输入，也可以通过 LAN 下载。

CNC 机床可以大大提高生产效率。然而，CNC 机床首先需要解决以下关键问题来确保高效率。

(1) 采购先进的 CNC 机床必须有足够的资金。

(2) 采购 CNC 机床必须与厂家签订全面服务合同，或设专职技术人员进行定期维护。

(3) 相关人员必须接受全面的 CNC 机床操作培训，尤其针对加工有几何公差的零件。

(4) 制订生产计划时必须符合 CNC 机床的实际情况，因为 CNC 机床每小时的运行成本通常高于普通机床每小时的运行成本。

1.4.2 CNC 机床的投资回报

鼓励企业将 CNC 机床视为一种生产解决方案，这具有以下好处。

(1) 节省直接劳动力。一台 CNC 机床的加工效率通常相当于多台普通机床的加工效率。

(2) 节省操作人员的培训成本。

(3) 节省车间管理成本。

(4) CNC 机床由于采用更严谨、可预测的生产计划而节省成本。

(5) 节省厂房占地面积，因为需要的 CNC 机床更少。

(6) 降低能耗，因为 CNC 机床加工零件的作业时间长，准备时间短。

(7) 简化成本估算和定价，从而节约成本。

(8) 不需要精密的工装夹具，减少了对专用工装夹具的需求，降低了工装夹具维护和存储成本。

(9) 减少刀具设计、制造和文档要求。CNC 机床减少了对特殊形状刀具、特殊镗杆、特殊螺纹刀具等的需求。

(10) 由于 CNC 机床能够加工高精度和重复性较高的零件，因此减少了检查时间。在许多时候，不影响作业时间的情况下，只需对关键工序进行抽检。

投资回收期用于估算投资效率。计算投资回收期时应考虑回收 CNC 机床净成本所需的年数，其计算公式为

$$投资回收期 = \frac{CNC 机床净成本 - CNC 机床净成本 \times 税收抵免率}{年收益额 - 年收益额 \times 税率 + 年计提折旧额 \times 税率} \tag{1.1}$$

投资回报率（return on investment，ROI）是指通过投资而应返回的价值，用于评估企业从某 CNC 机床投资中每年期望得到的经济回报率。计算投资回报率时应考虑 CNC 机床的使用寿命，其计算公式为

$$投资回报率 = \frac{年均利润 - CNC 机床净成本/使用年限}{CNC 机床净成本} \times 100\% \tag{1.2}$$

下面给出一个投资计算的例子（表 1-1），对新引进的 CNC 机床，确定其投资回收期和投资回报率。据保守估计，该 CNC 机床的使用寿命为 12 年。

表 1-1　CNC 机床投资回报表

期初投资（美元）	130250
一次性投资节省工装费用（美元）	35000
CNC 机床净成本（美元）	95250
年均利润（美元）	63100
税收抵免率（10%）	0.1
税率（46%）	0.46
CNC 机床年度折旧费（美元）	10900

$$投资回收期 = \frac{95250 - 95250 \times 0.1}{63100 - 63100 \times 0.46 + 10900 \times 0.46} \approx 2.19(年) \tag{1.3}$$

经过计算，该 CNC 机床的投资回收期约为 2.19 年。

$$投资回报率 = \frac{63100 - 95250 \div 12}{95250} \times 100\% \approx 57.9\% \tag{1.4}$$

按该 CNC 机床 12 年使用寿命计算，该 CNC 机床的投资回报率约为 57.9%（0.579×95250 美元 = 55149.75 美元）。

1.4.3　CNC 机床的可靠性

可靠性是指设备在规定的时间内及在规定的条件下完成预定功能的能力，当以概率来度量时，将其称为可靠度。产品主机的使用寿命可分为三个不同的时期。如图 1.17 所示，可靠性"浴盆曲线"描述了瞬时故障率与时间的关系。第一个时期称为早期失效期，第二个周期称为使用寿命期（正常寿命），第三个周期称为耗损失效期。从可靠性"浴盆曲线"爬升点开始，一直延伸到终点，零件因磨损、疲劳、老化和耗损，其故障率随时间的延长而急速增加。

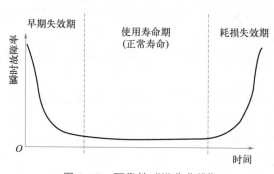

图 1.17　可靠性"浴盆曲线"

平均无故障工作时间（mean time between failures，MTBF）是衡量系统可靠性的指标，常用购买产品的每百万小时故障数表示，其在产品使用寿命和用户需求分析中很常见。MTBF 是系统在运行过程中两次故障间能正常工作的时间，可以描述为系统故障间隔的算术平均时间，其计算公式为

$$\text{MTBF} = T/R \tag{1.5}$$

式中，T——总工作时间；

R——总故障次数。

平均维修时间（mean time to repair，MTTR）是指系统从出现故障开始直到能正常工作所用的平均维修时间。在运行系统中，维修通常意味着要更换有故障的硬件。因此，硬件的平均维修时间常被视为更换故障硬件的平均时间。从长远来看，维修一个产品花费的时间太长，会增加安装成本。为了保证设备有较高的可靠性，许多企业会购买备件，以便产品出现故障时快速维修。

MTTR 是可维修项目的可维护性基本度量指标。它表示维修故障零件或设备所需的平均时间。用数学表示，它是给定的总维修时间与在该级别上被维修产品的总故障次数之比，其计算公式为

$$\mathrm{MTTR} = T/N \tag{1.6}$$

式中，T——总维修时间；
N——总故障次数。

本 章 小 结

本章讲解了数控技术的发展历史、现状及趋势，数控机床的基本原理，数控系统的分类及基本组成。

通过本章学习，要求学生掌握数控技术常用词汇，如 NC、CNC、DNC、point-to-point control system、straight-cut control system、contouring control system、interpolation、open-loop servo system、semi-closed loop servo system、closed-loop servo system、ROI、MTBF、MTTR 等；能够计算 CNC 机床的投资回收期和投资回报率。

本章的知识和能力图谱如图 1.18 所示。

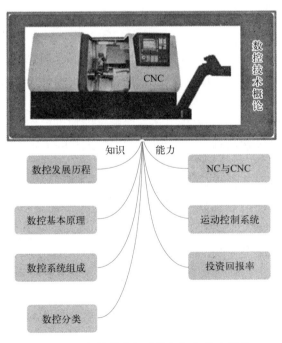

图 1.18 数控技术概论的知识和能力图谱

第2章 数控编程

教学目的及要求

  了解数控编程概述；
  了解数控编程基础；
  掌握数控编程约定；
  掌握数控编程实训；
  了解计算机辅助制造。

2.1 数控编程概述

  数控程序是一个编码指令列表，描述如何把所设计的零部件制造出来。这些编码指令称为数据，由一系列字母和数字组成。数控程序包括加工零件所有几何和技术要求，以及执行加工零件所需的机床功能和运动控制。

  数控程序可以分解成程序行，每行描述一组特定的加工操作，这些按顺序运行的行称为程序段。每个程序段包含功能字（代码），功能字表示特定的加工/移动指令或机床功能。

  数控装置能够识别的数控编程语言是一种 ISO 代码，它包括 G 代码和 M 代码。每个功能字都是一个控制机床的具体指令，它以一个英文字母开头，其后跟随一个或多个数字。

  零件数控程序可以包含多个单独的程序，这些程序一起描述加工零件所需的所有操作。主程序是主控程序，当运行整个加工程序时，系统首先读取和运行主程序指令。主程序可以调用较小的程序段。

  这些较小的程序段称为子程序，子程序可以被任何主程序或其他子程序调用，并且调用后自动返回调用的程序。通常，控制器只有一个主程序在运行，在这种情况下，主程序就是加工程序。主程序用 ISO 地址代码编写。

程序指令是指导并处理数控机床动作的具体步进命令。在数控机床中，程序指令称为加工程序，编制程序的人称为数控程序员。表示地址的英文字母、特殊文字和数字集合而成的编码指令实现对数控机床的自动控制。一系列用于执行操作的编码指令称为程序。程序被转换为相应的控制信号，输入数控机床，由伺服电动机控制动作。

数控为全程控制，从零件图到 CNC 加工的过程称为数控编程（CNC programming）。在用数控机床加工零件时，数控编程非常重要，数控编程不仅要正确、快速，而且要有效、经济。

2.1.1 数控编程的内容和步骤

数控编程前，数控程序员首先应了解数控机床的规格、特点，数控系统的功能，编程指令格式，等等；数控编程时，数控程序员首先应分析零件的技术要求、几何尺寸和工艺要求，然后确定加工方案，进行数值计算，得到刀具位置。根据零件尺寸、刀具位置、切削参数（主轴转速、进给速度、切削深度）和辅助功能（自动换刀、顺时针/逆时针旋转、冷却液开/关）等，完成编程。最后，数控程序员将程序输入数控系统，数控系统控制数控机床完成自动加工。

通常，数控程序员要遵循指定工作流程，这些流程可以概括为几个关键步骤：分析零件图并确定工艺流程，正确选择编程原点和坐标系，计算数值，编写程序，验证程序，编写程序文档并将程序输入 CNC 系统。

（1）分析零件图并确定工艺流程。

这一步骤包括分析零件图，了解加工内容和要求，确定工艺流程、加工规划、加工顺序、加工路线、夹紧方法、切削参数，选择合适的刀具，等等；除此之外，还应了解数控机床的代码格式，充分发挥数控机床的性能。

（2）正确选择编程原点和坐标系。

在数控加工中，正确选择编程原点和坐标系是非常重要的。在数控编程中，坐标系是指在工件上建立的坐标系。

（3）计算数值。

确定工艺流程后，根据零件的几何尺寸和刀具半径补偿方法，确定刀具轨迹，这样就可以得到刀具的位置坐标。

（4）编写程序。

数控程序员在确定工艺流程、加工路线和刀具坐标后，结合具体数控机床的程序格式，编写零件加工程序。

（5）验证程序。

在程序用于实际加工前，必须检查程序，检查可能的零件过切、夹具损坏、刀具破坏或使机床崩溃的错误刀具路径。在某些情况下，通过在数控机床上进行加工模拟来测试程序。在验证结果的基础上，对程序进行修改和调整，直至程序完全满足加工要求。

（6）编写程序文档并将程序输入 CNC 系统。

许多数控程序，不管它们是如何开发的，往往都缺乏帮助操作者了解程序的文档或注释信息，这是一个持续存在的问题。数控程序最好包括关于安装的基本信息，甚至专门的文档说明，这样操作者就能知道工装夹具的种类、零件的定位方向、刀具的种类及零件所在的位置。

程序编码通过合适的控制介质提交给 MCU。过去，常用的控制介质是穿孔带，使用标准格式，由 MCU 读取。现在，在加工车间，穿孔带基本上已被其他存储技术所取代，这些技术包括磁带、磁盘、USB 闪存驱动器和计算机直接传输等。

以上所述是人工编程步骤，即手工编程。一个数控程序员不仅要了解数控机床的结构、数控系统的功能和标准，还必须了解工艺知识，如工装夹具、刀具和切削参数等。

要特别提醒的是，以上步骤并不是必须按部就班地完成。通常，每个编程步骤都是相关联的，编程往往会推倒重来，重新编写。

一旦零件图和程序到达加工车间，数控机床操作员就开始现场加工。加工循环开始前，还要完成以下工作流程。

（1）评价程序。

（2）检查毛坯及材料。

（3）准备刀具。

（4）安装和设置刀具。

（5）安装工件。

（6）下载程序。

（7）设置刀具补偿。

（8）加工首件。

（9）优化程序（若有必要）。

（10）加工工件。

（11）不定时检查程序。

2.1.2 数控编程方法

数控编程方法很多，包括手工编程和自动编程。

1. 手工编程

手工编程是指整个数控编程过程（包括在计算机上的数值计算）由人工完成。

在机械加工中，许多简单零件由简单的直线和圆弧等几何要素组成。零件的数值计算简单，程序段不多，程序检查容易，这些零件程序可以手动完成。因此，手工编程仍然是国内外普遍应用的编程方法。

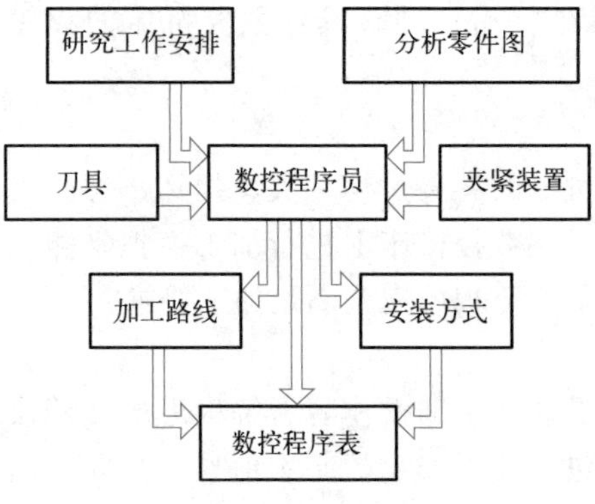

如图 2.1 所示，手工编程可以归纳为以下四个步骤。

（1）分析零件图。

（2）确定加工路线。

（3）选择夹具和刀具（数控安装单）。

（4）编写数控程序（数控程序表）。

数控编程时，必须仔细研读相关文档，并严格制订生产执行计划。

因为手工编程对于具有非圆曲线和曲面的复

图 2.1 手工编程

杂零件的编程有困难或难以实现，所以必须使用自动编程来实现。

2. 自动编程

自动编程也称计算机辅助编程，是指由计算机完成数控加工程序编制过程中的全部或大部分工作，由计算机系统完成大量的数值计算、程序编写和控制介质制作等。自动编程可以减轻数控程序员的劳动强度，大大提高编程效率和编程质量。同时，自动编程能够顺利完成一些计算烦琐、手工编程困难或无法编出的程序。零件的形状和工艺越复杂，相对手工编程，自动编程的优势越明显。

自动编程可以分为以自动编程语言为基础的 APT（automatically programmed tools）和会话式 APT 两类。自动编程语言为基础的 APT 在编程时，数控程序员依据所用数控语言的编程手册，以及加工顺序、零件图样和刀具参数，以语言的形式表达加工的全部内容，然后把这些内容全部输入计算机进行处理，制作出可以直接用于数控机床的加工程序。会话式 APT 在编程时，数控程序员首先通过图形用户界面（graphical user interface, GUI）对零件图进行交互输入，然后确定加工顺序，输入相应的切削加工工艺参数，最后 CNC 系统基于输入数据完成加工程序。典型的会话式 APT 由外部计算机辅助制造系统和位于 CNC 系统内部或计算机中的图示化对话系统执行。

2.2 数控编程基础

2.2.1 数控机床的坐标系统

用来确定机床运动位置的 X、Y、Z 坐标系称为笛卡儿坐标系，笛卡儿坐标系包括三条数轴，相互成 90°相交，相交点称为原点。

几乎所有能够在普通机床上加工的零件都适合在数控机床上加工。按数控机床运动轨迹，可将数控机床的控制方式分为点到点控制、直线切削控制和轮廓控制（连续控制）。

笛卡儿坐标系是法国数学家、哲学家和物理学家勒内·笛卡儿发明的，通过坐标系建立了任何一点的坐标和三条数轴的数学模型，如图 2.2 所示。数控机床的运动采用笛卡儿坐标系来描述，包括在切削加工过程中 X、Y、Z 轴移动和绕 X、Y、Z 轴的转动。在普通立式铣床上，X 轴是工作台的水平运动（右或左），Y 轴是工作台的横向运动（前或后），Z 轴是升降台或主轴的上下运动。数控系统依赖于机床坐标系统，因为数控程序员在编程时依赖笛卡儿坐标系确定工件上每个点的位置。

通常，数控机床运动包括 6 个自由度，如图 2.3 所示，包括沿 X、Y、Z 三直线轴移动（X、Y、Z）和绕 X、Y、Z 轴的转动（A、B、C）。

数控机床要求加工工件时进给运动和各回转轴运动可控，进给伺服系统能快速调节各独立坐标轴的运动速度，并能精确地进行位置控制，能够在各伺服轴上相互独立地控制进给运动，即多轴联动。现代数控机床基本淘汰了普通机床的各种操作手柄和手轮。

数控车床至少可以同时控制协调两个坐标轴联动（X 轴和 Z 轴），如图 2.4 所示。数控

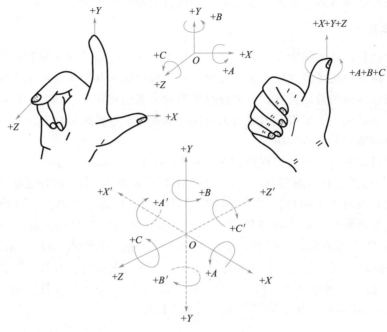

图 2.2 笛卡儿坐标系

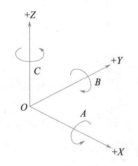

图 2.3 数控机床的三轴
直线移动和转动

铣床工件夹紧装置用于固定工件,使之占有正确的位置,其坐标轴如图 2.5 所示。夹具要求装夹方便、迅速,定位准确、可靠,重复性好。多数的数控加工采用液压夹紧,基本满足工件夹紧力要求。对于铣削而言,加工过程要尽可能地减少重新装夹次数。机床坐标系和工件坐标系分别如图 2.6 和图 2.7 所示。

因为铣削复杂回转体零件时一次装夹完成多工序加工,所以建议使用自动化装置。例如,托盘交换装置被越来越多地使用,它由操作员在数控机床外装夹下一个工件,然后自动推入机床的加工位置。

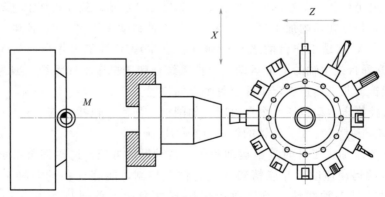

图 2.4 数控车床坐标轴

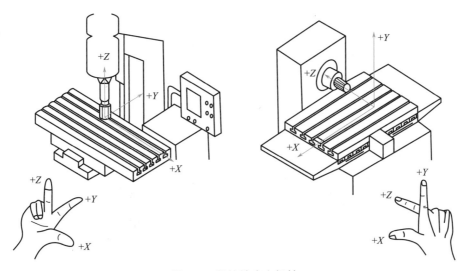

图 2.5 数控铣床坐标轴

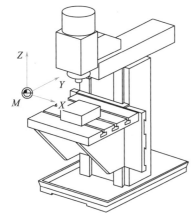

图 2.6 机床坐标系

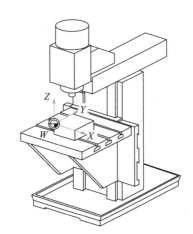

图 2.7 工件坐标系

2.2.2 坐标系

1. 绝对坐标系

绝对坐标系是指所有坐标全部基于一个固定坐标系原点的位置来描述坐标。刀具（或机床）运动位置的所有指令坐标值都是相对于固定的坐标原点给出的。

2. 增量坐标系

增量坐标系是指相对于前一坐标点的位置来描述坐标。其缺点是坐标原点经常变换，一旦出错，将会出现累积误差。

2.2.3 原点和参考点

数控铣床原点和参考点示意图如图 2.8 所示。

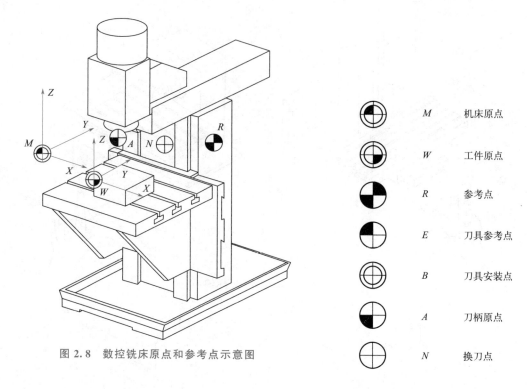

图 2.8　数控铣床原点和参考点示意图

1. 机床原点 M

每台数控机床都有一个机床坐标系。机床零点是机床坐标系的原点，它在机械硬件上的位置由制造厂家确定，是机床上的一个固定点。通常，数控车床的机床原点设在卡盘端面与主轴中心线交点处，而数控立式铣床的机床原点设在机床 X、Y、Z 三轴正方向的运动极限位置处。

2. 工件原点 W（编程原点）

工件原点应选在零件图的基准上，它是工件的基准点。数控程序员确定工件原点（编程原点）时通常要考虑利于编程、便于测量等因素。例如，工件原点可以选在工件外轮廓的某一角上，对于对称的工件，最好选在工件的对称中心上。无论工件原点选在何处，机床操作员都要获得工件原点信息，加工开始时，在手动状态下通过对刀等方式将工件原点与数控机床坐标建立联系，以便刀具简便且准确地定位。数控程序员通过工件坐标系建立刀位与机床原点的坐标关系，操作员将工件原点的坐标值输入 MCU。

3. 参考点 R

数控装置上电时并不知道机床原点的位置，机床工作时为了正确建立机床坐标系，通常在每个坐标轴的移动范围内设置一个机床参考点 R（测量起点），机床启动时通常要进行自动或手动返回参考点，以建立机床坐标系并激活参数。数控机床加工时，操作员还要搞懂工件坐标系的定义及如何实现对刀。

增量式测量系统的数控机床需要一个测量起点，并用于刀具和工件的运动控制，这个点称为机床参考点 R。机床参考点的位置由制造厂家在每个进给轴上用限位开关精确调整好,坐标值已输入数控系统。因此机床参考点对机床原点的坐标是一个已知数。数控机床开机时,必须先确定机床原点,即刀架返回参考点的操作;只有机床参考点被确认后,刀具(或工作台)移动才有基准。

4. 刀位点

在机床外测量刀具时,刀具、刀架(安装)和刀柄上的刀具定位基准点必须确定,因为控制系统需要将刀具的几何信息(如长度和半径)与该基准点建立联系,以便将加工程序的刀具坐标值精确地应用到工件上。换刀应在安全的位置(换刀点)进行。换刀点在加工工件的轮廓外,并留有一定的安全空间。

2.3 数控编程约定

数控编程是将所有的工艺数据用数控装置所规定的规则指令和格式编成加工程序单的过程。工艺数据的内容如下。

(1) 加工方法和工艺顺序,如刀具切入点、切削深度、刀具轨迹等。
(2) 工艺参数,如主轴转速、进给量、冷却液开/关等。
(3) 切削刀具选择。

数控程序由程序段、功能字和地址组成,如图2.9所示。

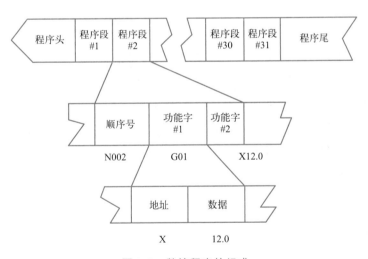

图 2.9 数控程序的组成

(1) 程序段——实现一个完整的加工工步或动作。
(2) 功能字——程序段由若干功能字组成,每个功能字由地址符和数字组成,如F200表示进给速度为200mm/min。
(3) 地址——每个字都由字母开头,见表2-1。

表 2-1 地址

地址	代码/规则	功能
% MM	ISO	子程序
% PM	ISO	主程序号
% TM	ISO	刀具补偿表
N	9001—9999999	程序号和子程序号
N	1—8999	程序段号
X	+/− 9999,99	X 坐标尺寸字（mm）
Y	+/− 9999,99	Y 坐标尺寸字（mm）
Z	+/− 9999,99	Z 坐标尺寸字（mm）
B	+/− 9999,99	B 坐标尺寸字（°）
R	+/− 9999,99	圆弧半径（mm）
I	+/− 9999,99	圆弧中心 X 坐标
J	+/− 9999,99	圆弧中心 Y 坐标
K	+/− 9999,99	圆弧中心 Z 坐标
P	0—9999	子程序调用次数
F	1—5000	进给速度（mm/r 或 mm/min）
S	20—99999	主轴转速（r/min）
T	0—99	刀具补偿号
E	0—99	子程序参数

一般程序段（程序行）由下列功能字按顺序组成。

N＿＿G＿＿X＿＿Y＿＿Z＿＿F＿＿M＿＿S＿＿T＿＿；

在数控程序中，每个程序段的功能字顺序都是相同的，但具体到某个程序段，不一定必须包括以上所有功能字。

数控程序例子如下。

N20 G01 X20.5 F200 S1000 M03;
N21 G02 X30.0 Y40.0 I20.5 J32.0;

1. 程序段号（N）

程序段号也称顺序号字，用以标识程序段。它位于程序段之首，相当于该程序段名。需要注意的是，数控程序是按程序段的排列次序执行的，与顺序段号的大小、次序无关，即程序段号实际上只是程序段的名称，而不是程序段执行的先后次序。实际上，有些数控系统并不一定需要程序段号。

2. 准备功能字（G 代码）

准备功能字常称 G 代码，它是控制刀具运动的主要功能类别。常见的 G 代码及其功能（数控车床）见表2-2。

表 2-2 常见的 G 代码及其功能（数控车床）

G 代码	功能
G00*	点定位（快速运动）
G01	直线插补
G02	顺时针（CW）圆弧插补
G03	逆时针（CCW）圆弧插补
G04	暂停
G20	英制输入（英寸）
G21*	公制输入（毫米）
G22*	行程检查开关（开）
G23	行程检查开关（关）
G27	参考点返回检查
G28	自动返回参考点
G29	从参考点返回
G30	从第二参考点返回
G32*	螺纹切削
G40	刀具半径补偿/刀具偏置的取消
G41	刀具半径补偿（左）
G42	刀具半径补偿（右）
G50	工件坐标系设定/主轴最大速度限定
G70	精车固定循环
G71	内外径粗车复合循环
G72	端面粗车循环
G73	固定形状粗车循环
G74	端面沟槽循环
G75	外圆切槽循环
G76	多头螺纹切削循环
G90	绝对坐标编程
G91	增量坐标编程
G92	螺纹切削循环
G94	端面切削循环
G96	恒线速度控制（mm/min）
G97*	恒线速度控制取消（r/min）
G98	进给速度：每分钟进给量（mm/min）
G99*	进给速度：每转进给量（mm/r）

注：1. "*"表示模态指令，模态指令又称续效指令，一经程序段指定，便一直有效，直到后面出现同组另一指令或被其他指令取消时才失效。

2. 通常，G代码用于数控车床，可以通过参数设定使其具有特殊用途。

3. 圆弧插补指令参数（I/J/K）

这些参数指定了圆弧终点相对于圆弧起点的坐标值和圆心坐标值。I＿ J＿ K＿的坐标值分别对应于 X＿ Y＿ Z＿的坐标。

4. 主轴功能字（S）

主轴功能字用来指定主轴速度，其单位为 r/min，主轴转速的计算公式为

$$主轴转速 = \frac{切削速度 \times 1000}{\pi \times 刀具直径} \tag{2.1}$$

5. 进给功能字（F）

进给功能字用来指定除快速运动指令外刀具相对工件运动的速度，其单位为 mm/min（一般为数控铣床）或 mm/r（一般为数控车床）。选用何种单位必须在编程开始时就确定，进给速度的计算公式为

$$进给速度 = 每齿进给量 \times 刀齿数 \times 刀具转速 \tag{2.2}$$

6. 辅助功能字（M 代码）

辅助功能字常称 M 代码，主要用于数控机床的开关量控制，一般与刀具运动无关。常见的 M 代码及其功能见表 2-3。

表 2-3 常见的 M 代码及其功能

M 代码	功能
M00	程序停止
M02	程序结束
M03	主轴顺时针旋转（CW）
M04	主轴逆时针旋转（CCW）
M05	主轴停止
M06	刀具交换
M07	内冷却液开
M08	外冷却液开
M09	冷却液关
M10	夹紧
M11	松开
M12	尾架轴伸出
M13	尾架轴缩进
M19	主轴定向
M30	程序结束，程序执行光标回到程序头

续表

M 代码	功能
M66	手动换刀
M67	假换刀
M98	子程序调用
M99	子程序返回

2.4 数控编程实训

数控程序中包含许多功能字,如 G 代码(准备功能字)和 M 代码(辅助功能字),它们是数控编程的基础。国际标准化组织(ISO)制定了 G 代码和 M 代码的参考标准。由于新的数控系统和机床不断涌现,它们的代码更加丰富,格式也更加灵活,超出了 ISO 标准的限制,即使是同一功能,其代码和格式在不同制造厂家开发的系统间也有很大的差异,甚至同一制造厂家生产的新系统和旧系统之间代码和格式也不一定相同。但是,在大多数 CNC 系统中,G 代码和 M 代码都参考 ISO 标准。

通常,加工一个工件需要多把刀具,每把刀具的长度不同,如图 2.10 所示。

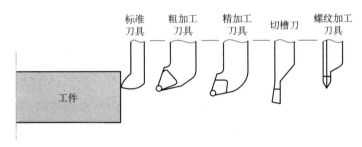

图 2.10 不同刀具长度

因此,在编程前要考虑每把刀具的长度设置,一般在数控系统中选择一把刀具作为基准刀,用基准刀对刀,将其他刀具相对于基准刀的偏移值设置在各自的刀具偏差补偿中。这样,更换非基准刀后无须修改程序,该功能称为刀具长度补偿。

2.4.1 车削加工中心 MAHO GR350C 编程

MAHO GR350C 的操作面板如图 2.11 所示。

程序示例(MAHO GR350C):

```
% PM1234567      (% PM+ N   N 为程序号)
N1234567         (程序体)
N1 G18;
N2 G52;
…
N2000 M30;       (程序结束)
```

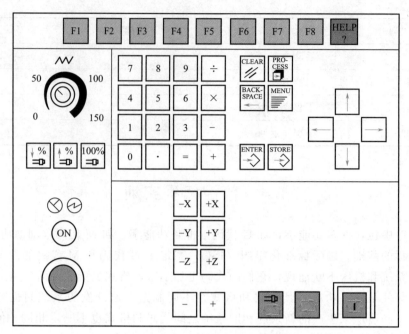

图 2.11 MAHO GR350C 的操作面板

1. G10：轴向容积切削加工循环

指令格式：G10 X__(U__)Z__(W__)I__K__C__(F__)

其中，X——起始点 S 的 X 绝对坐标；

U——起始点 S 的 X 增量坐标；

Z——起始点 S 的 Z 绝对坐标；

W——起始点 S 的 Z 增量坐标；

I——X 方向为 G12 精加工循环所保留的精加工余量；

K——Z 方向为 G12 精加工循环所保留的精加工余量；

C——每次的切削深度；

F——进给量，单位通常为 mm/r。

G10 实例（图 2.12）：

% PM241001

N241001

N1 G54;   (定义工件零点)

N2 G99 X140 Z127;   (毛坯定义)

N3 G96 S100 D2500 T1013 M4;   (恒线速度切削选择,切削线速度为100m/min,主轴最高转速不超过 2500r/min)

N4 G10 X145 Z130 I0.5 K0.5 C2.5 F0.5; (循环定义,精加工余量 X、Z 方向均为 0.5mm,每次切削厚度为 2.5mm,轮廓起点为 145、130)

N8001 G01 X40 Z125;   (轮廓定义开始)

N5 G01 W-5;

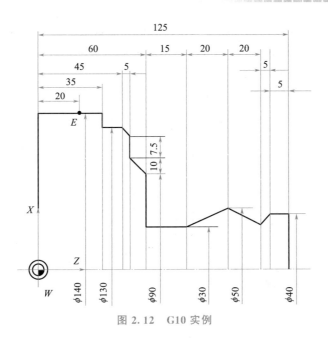

图 2.12　G10 实例

N6 G01 X30 W-5;

N7 G01 X50 Z95;

N8 G01 X30 Z75;

N9 G01 Z60;

N10 G01 X90;

N11 G01 U20 Z50;

N12 G01 U15;

N13 G01 X130 Z45;

N14 G01 Z35;

N15 G01 X140;

N16 G01 Z28;　　　　　　　　　　　　（轮廓定义结束）

N17 G13 N1= 8001 N2= 16;　　　　　　（容积切削加工循环调用）

N18 G00 X200 Z145;　　　　　　　　　（退刀）

N19 T2023;　　　　　　　　　　　　　（换刀）

N20 G12 X145 Z127 S200;　　　　　　 （轮廓精加工循环）

N21 G13 N1= 8001 N2= 16;　　　　　　（容积切削加工循环调用）

N22 G00 X300 Z350;　　　　　　　　　（退刀）

N23 M30;

2. G11：径向容积切削加工循环（图 2.13）

指令格式：G11 X＿(U＿)Z＿(W＿)I＿K＿C＿(F＿)

其中，X——起始点 S 的 X 绝对坐标；

U——起始点 S 的 X 增量坐标；

Z——起始点 S 的 Z 绝对坐标；

W——起始点 S 的 Z 增量坐标；

I——X 方向为 G12 精加工循环所保留的精加工余量；
K——Z 方向为 G12 精加工循环所保留的精加工余量；
C——每次切削深度；
F——进给量，单位通常为 mm/r。

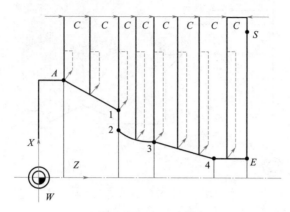

图 2.13　G11 功能

G11 实例（图 2.14）：

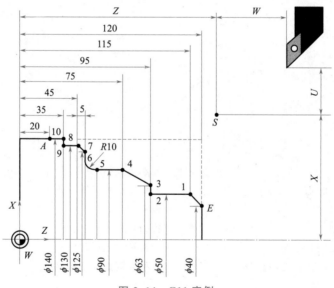

图 2.14　G11 实例

```
% PM241002
N241002
N1 G54;                          （工件坐标系）
N2 G99 X140 Z127;                （毛坯定义）
N3 G96 S100 D2500 T1013 M4;      （恒线速度切削选择，切削线速度为 100m/min, 主轴最高转速不超过
                                  2500r/min）
```

N4 G00 X200 Z145;　　　　　　　　（快速移动到起始点）
N5 G11 X145 Z120 I0.5 K0.5 C2.5 F0.5;
（循环定义,精加工余量X、Z方向均为0.5mm,每次切削厚度为2.5mm,轮廓起点为145、120）
N6 G01 Z25;　　　　　　　　　　　（轮廓定义开始）
N7 G41 X140;　　　　　　　　　　　（刀具半径补偿:左）
N8 G01 Z35;
N9 G01 X130;
N10 G01 Z45 F0.05;
N11 G01 X125 Z50;
N12 G01 X110;
N13 G03 X90 Z60 R10;
N14 G01 Z75;
N15 G01 X63 Z95 F0.1;
N16 G01 X50;
N17 G01 Z115;
N18 G01 X40 Z120;　　　　　　　　（轮廓定义结束）
N19 G40;　　　　　　　　　　　　　（取消刀具半径补偿）
N20 G13 N1= 6 N2= 19;　　　　　　（容积切削加工循环调用）
N21 G00 X200 Z145;　　　　　　　　（退刀）
N22 T2023;　　　　　　　　　　　　（换刀）
N23 G12 X145 Z120 S200;　　　　　（轮廓精加工循环）
N24 G13 N1= 6 N2= 19;　　　　　　（容积切削加工循环调用）
N25 G00 X300 Z350;
N26 M30;

3. G12：轮廓精加工循环

指令格式：G12 X __(U __)Z __(W __)(F __)

其中，X——起始点S的X绝对坐标；
　　　U——起始点S的X增量坐标；
　　　Z——起始点S的Z绝对坐标；
　　　W——起始点S的Z增量坐标。

4. G13：容积切削加工循环调用（执行）

指令格式：G13 N1= __ N2= __

其中，N1——轮廓定义的第一条语句；
　　　N2——轮廓定义的最后一条语句。

5. G96：选择恒线速度切削与G97：取消恒线速度切削

指令格式：G96 S __ F __ D __

其中，S——切削线速度（m/min）；
　　　F——进给速度；
　　　D——主轴最高转速（r/min）。

指令格式：G97 S＿＿

其中，S——主轴最高转速（r/min）。

6. G32：螺纹车削加工循环

指令格式：G32 X＿＿(U＿＿)Z＿＿(W＿＿)C＿＿(D＿＿)(A＿＿)(J＿＿)(B＿＿)F＿＿

其中，X——螺纹底径；

　　　　Z——Z 坐标终点，相对编程零点；

　　　　U——螺纹深度（从起始点算起），＋U 内螺纹，－U 外螺纹；

　　　　W——螺纹长度（从起始点算起），可用±号；

　　　　C——第一次切削深度，如果 C＝U 则切削一次，如果 C＜U 则切削多次；

　　　　D——螺纹精加工深度；

　　　　A——刀尖角的一半；

　　　　J——螺纹切出锥度的终止直径；

　　　　B——锥度比例；

　　　　F——螺纹螺距。

G32 参数如图 2.15 所示。

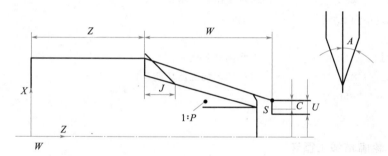

图 2.15　G32 参数

2.4.2　FANUC 0i MATE TB 数控车削编程

FANUC 0i MATE TB 数控系统 G 代码及其功能见表 2-4。

表 2-4　FANUC 0i MATE TB 数控系统 G 代码及其功能

G 代码	功能	G 代码	功能
G00	点定位（快速运动）	G12	英制输入（英寸）
G01	直线插补（直线切削）	G13	公制输入（毫米）
G02	圆弧插补（顺时针）	G27	自动返回参考确认
G03	圆弧插补（逆时针）	G28	返回参考位置
G04	暂停（延时）	G31	跳转指令
G10	可编程数据输入（资料设定）	G32	螺纹切削
G11	取消可编程数据输入	G40	取消刀具（刀尖）半径补偿

续表

G 代码	功能	G 代码	功能
G41	刀具（刀尖）半径左补偿	G70	精车固定循环
G42	刀具（刀尖）半径右补偿	G71	外圆粗车循环
G50	工件坐标系设定或主轴最大速度设定	G72	端面粗车循环
G52	局部坐标系设定	G73	固定形状粗车循环
G53	机床坐标系设定	G74	端面沟槽循环
G54	工件坐标系选择 1	G75	外径/孔断续切槽循环
G55	工件坐标系选择 2	G76	多头螺纹切削循环
G56	工件坐标系选择 3	G90	外圆/孔切削循环
G57	工件坐标系选择 4	G92	螺纹切削循环
G58	工件坐标系选择 5	G94	端面切削循环
G59	工件坐标系选择 6	G96	恒线速度控制
G65	调用宏指定	G97	恒线速度控制取消
G66	模态宏调用	G98	进给速度：每分钟进给量（mm/min）
G67	注销模态宏调用	G99	进给速度：每转进给量（mm/r）

1. G01：直线插补（切削进给）

指令格式：G01 X(U)__ Z(W)__ F__

G01 实例（图 2.16）：

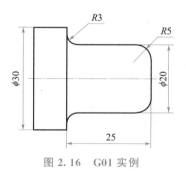

图 2.16　G01 实例

O2421
N10 T0101;
N20 G0 X0 Z1.S1000 M03;　（绝对坐标）
N30 G1 Z0.F0.2;
N40 G1 X20.R-5.;
N50 G1 Z-25.R3.;
N60 G1 X30.5;

N70 G28 X120. Z100. ;
N80 M30;

2. G04：暂停

指令格式：G04 X(U)__ or G04 P__

其中，X——延时（其后需加小数点）；
  U——延时（其后需加小数点）；
  P——延时（整数，无小数点）。

3. G28：返回参考位置

指令格式：G28 X(U)__ Z(W)__

该指令首先用快速进给，使各轴移向中间点（中间点为程序执行前坐标），然后从该中间点向参考点进行快速进给定位，最后返回参考点。为安全起见，在执行该指令前，应取消刀具半径补偿和偏置。

4. G32：螺纹切削

指令格式：G32 X(U)__ Z(W)__ F__

螺纹切削通常从粗车到精车需要刀具多次在同一轨迹上进行切削。由于螺纹切削是从检测出主轴上的位置编码器一转信号后开始的，因此，无论进行几次螺纹切削，工件圆周上切削起点都是相同的，螺纹切削轨迹也都是相同的。但是，从粗车到精车主轴的转速必须是恒定的，主轴转速发生变化时，螺纹切削会产生一些偏差。

G32 实例（图 2.17）：

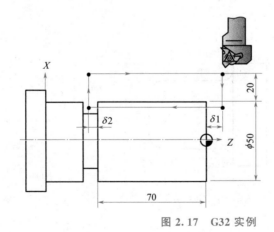

图 2.17 G32 实例

O2422
N10 G50 T0101;
N20 G97 S1000 M03;
N30 G00 X90.0 Z5.0 T0202 M8;
N40 X48.0;
N50 G32 Z-71.5 F3.0;

N60 G00 X90.0;
N70 Z5.0;
N80 X46.0;
N90 G32 Z-71.5;
N100 G00 X90.0;
N110 Z5.0
N120 X150.0 Z150.0 T0101;
N130 M30;

5. G50：工件坐标系设定或主轴最大转速设定

G50 无参数，设定工件坐标系或主轴最大转速。

6. G90：绝对坐标指令与 G91：增量坐标指：

指令格式：G90 X(U)__ Z(W)__ F__

各轴移动量的指令方法有绝对坐标指令与增量坐标指令两种。绝对坐标指令以各轴移动的终点坐标值编程，增量坐标指令以各轴运动的位移量编程。

G90 实例 1（图 2.18）：

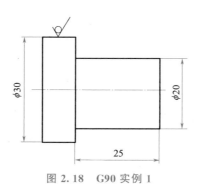

图 2.18　G90 实例 1

O2423
N10 T0101;
N20 G0 X31. Z1. S800 M03;   (快速走刀至循环起点)
N30 G90 X26. Z-24.9 F0.3;   (X 方向切深单边量为 2mm，端面留余量为 0.1mm，精加工)
N40 X22.;
N50 X20.5;   (X 向单边余量为 0.25mm)
N70 X20. Z-25. F0.2 S1200;   (精车)
N80 G28 X100. Z100.;
N90 M30;

G90 实例 2（图 2.19）：

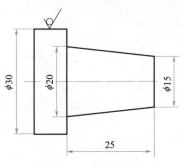

图 2.19　G90 实例 2

```
O2424
N10 T0101;                        (刀具)
N20 G0 X32. Z0.5 S1000 M3;
N30 G90 X26. Z-25. R-2.5 F0.15;   (粗加工)
N40 X22.;
N50 X20.5;                        (留精加工双边余量为 0.5mm)
N60 G0 Z0 S1500 M3;
N70 G90 X20. Z-25. R-2.5 F0.1;
N80 G28 X100. Z100.;
N90 M5;
N100 M2;
```

7. G92：螺纹切削循环

指令格式：G92 X(U)__ Z(W)__ F__

螺纹导程范围及主轴转速等参数与 G32 螺纹切削相同。该循环指令可以完成螺纹倒角。当到达某一位置时，接收从机床来的信号，启动螺纹倒角。螺纹倒角距离在 (0.1～12.7)L 指定，指定单位为 0.1L，由参数♯5130 决定。（上述描述中，L 为螺纹导程：单头螺纹导程等于螺距，多头螺纹导程等于螺纹头数乘以螺距）。

G92 实例 1（图 2.20）：

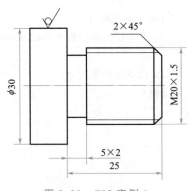

图 2.20　G92 实例 1

O2425
N110 T0303;
N120 G0 X28. Z5. S350 M3;   (刀具定位)
N130 G92 X19.4 Z-23. F1.5;   (螺纹加工)
N140 X19.;   (逐层进刀)
N150 X18.6;
N160 X18.2;
N170 X18.;
N180 X17.9;
N190 X17.8;
…

指令格式：G92 X(U)__ Z(W)__ R__ F__

其中，F——指定螺纹导程 L；

R——被加工锥形螺纹的大小端外径差的二分之一，即单边锥度差值。对外螺纹车削，锥度左大右小，R 值为负；反之为正。对内螺纹车削，锥度左小右大，R 值为正；反之为负。

G92 实例 2（$P=1.5$，图 2.21）：

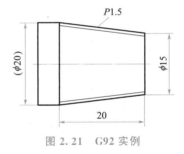

图 2.21 G92 实例

O2426
N10 T0101;
N20 G0 X25. Z5. S300 M3;
N30 G92 X19.6 Z-20. R-2.5 F2;
N40 X19.4;
N50 X19.;
…

8. G94：端面切削循环

指令格式：G94 X(U)__ Z(W)__ F__

G94 参数如图 2.22 所示。

当以增量坐标指令时，地址 U、W 的符号由轨迹 1 及轨迹 2 的方向决定。如果是 Z 轴的负方向，则 W 的数值为负。

单程序段时，用一次循环，可进行 1、2、3、4 的操作。

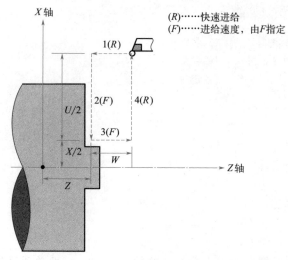

图 2.22 G94 参数

G94 也可用于锥面加工循环。

G94 实例（图 2.23）：

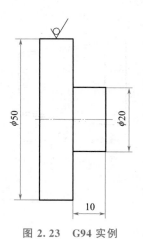

图 2.23 G94 实例

O2427
N10 T0101;
N20 G0 X52. Z1. S1000 M03;
N30 G94 X20.2 Z-2. F0.2;   （粗车第一刀，Z 向切削深度为 2mm）
N40 Z-4.;
N50 Z-6.;
N60 Z-8.;
N70 Z-9.8;
N80 X20. Z-10. S1500;   （精加工）
N90 G28 X100. Z100.;

```
N100 M30；
```

9. G70：精车固定循环

指令格式：G70 P(ns)__ Q(nf)__

其中，(ns)——精加工描述程序的开始循环程序段的行号；

(nf)——精加工描述程序的结束循环程序段的行号。

G70 指令用于在 G71、G72、G73 指令粗车工件后，进行精车固定循环。

在 G70 状态下，在指定的精车描述程序段中 ns~nf 的 F、S、T 有效。若不指定精车描述程序段，则维持粗车前指定的 F、S、T 状态。

当 G70 循环结束时，刀具返回起点并读取下一个程序段。

G70~G73 中在 ns 与 nf 间的程序段不能调用子程序。

10. G71：外径粗车循环

指令格式：G71 U(Δd)__ R(e)__
　　　　　　G71 P(ns)__ Q(nf)__ U(Δu)__ W(Δw)__ F__ S__ T__

其中，Δd——循环每次的切削深度（半径值、正值），该值为无符号值，切削方向由 AA'
　　　　方向指定，该值是模态值，在下次指定之前一直有效，它由参数♯5132 指
　　　　定，通过程序指令修改；

　　　e——每次切削退刀量，该值是模态值，在下次指定之前一直有效，它由参数
　　　　♯5133 指定，通过程序指令修改；

　　　ns——精加工描述程序时开始循环程序段的行号；

　　　nf——精加工描述程序时结束循环程序段的行号；

　　　Δu——X 向精车预留量（直径/半径）；

　　　Δw——Z 向精车预留量；

　　　F，S，T——在 ns 与 nf 间的运动指令中指定的 F、S、T 功能对粗加工循环无效，
　　　　　　　　对精加工有效，在 G71 程序段或前面程序段中指定的 F、S、T 功能
　　　　　　　　对粗加工有效。

最终轮廓 $A—A'—B$ 由程序给定，G71 参数如图 2.24 所示。指定区域以 Δd（切削深度）精车，预留量 Δu/2 和 Δw 被加工掉。

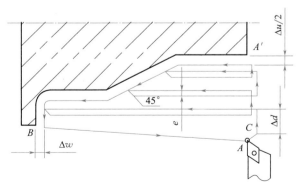

图 2.24　G71 参数

G71 实例（图 2.25）：

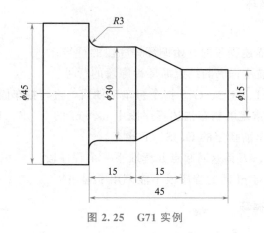

图 2.25 G71 实例

O2428
N10 T0101;
N20 G0 X46. Z0.5 S1000 M03;
N30 G71 U2. R0.5;   （每层切削深度为 2mm,退刀 0.5mm）
N40 G71 P50 Q110 U0.3 W0.1 F0.3; （精加工余量 X 向单边量为 0.3mm,Z 向单边量为 0.1mm,粗车进给量为 0.3mm/r）
N50 G1 X15.;
N60 G1 Z0. F0.15 S1500;   （精加工进给量为 0.15mm/r,精车主轴转速为 800r/min）
N70 Z-15.;
N80 X30. Z-30.;
N90 Z-42.;
N100 G2 X36. Z-45. R3.;
N110 G1 X46.;
N120 G70 P50 Q100;   （精加工循环）
N130 G28 X100. Z100.;
N140 M5;
N150 M30;

11. G72：端面粗车循环

指令格式：G72 W(d)__ R(e)__
　　　　　G72 P(ns)__ Q(nf)__ U(u)__ W(w)__ F__ S__ T__

G72 参数如图 2.26 所示，G72 指令的含义与 G71 类似，不同之处是刀具平行于 X 轴方向切削，该循环方式适用于对长径比较小的盘类工件端面方向粗车。

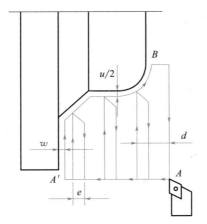

图 2.26 G72 参数

G72 实例（图 2.27）：

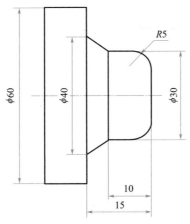

图 2.27 G72 实例

O2429

N10 T0101;

N20 G0 X61. Z0.5 S1000 M03;

N30 G72 W2. R0.5;

N40 G72 P50 Q100 U0.1 W0.3 F0.25;

N50 G0 Z-15.;

N60 G1 X40. F0.15 S1500;

N70 X30. Z-10.;

N80 Z-5.;

N90 G2 X20. Z0 R5.;

N100 G0 Z0.5;

N110 G70 P60 Q110;

N120 G28 X100. Z100.;

N130 M30;

12. G73：固定形状粗车循环

指令格式：G73 U(Δi)__ W(Δk)__ R(Δd)__
　　　　　G73 P(ns)__ Q(nf)__ U(Δu)__ W(Δw)__ F__ S__ T__

固定形状粗车循环也称成形加工复合循环，只要编出最终加工路线，给出每次切削的余量深度或循环次数，机床就可自动重复切削，直到工件加工完为止。它适用于加工形状已经基本成形的零件或毛坯尺寸接近工件成品尺寸的零件，如铸造、锻造毛坯零件等。

G73 实例（图 2.28）：

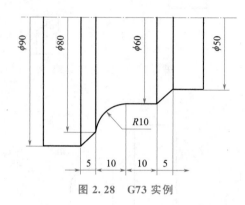

图 2.28　G73 实例

```
O24210
N10 T0101;
N20 G0 X110.Z10.S800 M3;
N30 G73 U5.W3.R3.;
N40 G73 P50 Q110 U0.4 W0.1 F0.3;
N50 G0 X50.Z1.S1200;
N60 G1 Z-10.F0.15;
N70 X60.Z-15.;
N80 Z-25.;
N90 G2 X80.Z-35.R10.;
N100 G1 X90.Z-40.;
N110 G0 X110.Z10.;
N120 G70 P50 Q110;
N130 G28 X100.Z150.;
N140 M30;
```

13. G74：端面沟槽循环

指令格式：G74 R(e)__
　　　　　G74 X(u)__ Z(w)__ P(Δi)__ Q(Δk)__ R(Δd)__ F__

其中，e——每次啄式退刀量，该值是模态值，在下次指定之前一直有效，它由参数 ♯5139 指定，通过程序指令修改；

　　X——X 向终点 B 绝对坐标；

　　U——从 A 到 B 增量值（X 坐标）；

Z——Z 向终点 C 绝对坐标；

W——从 A 到 C 增量值（Z 坐标）；

Δi——X 向每次的移动量（无符号值），1000 表示 1mm；

Δk——Z 向每次的切入量（无符号值）；

Δd——切削到终点（槽底）时的 X 轴退刀量（可以省略）用正值确定，如果省略 X（U）和 Δi，则需指定退刀方向的符号；

F——进给速度。

G74 实例（图 2.29）：

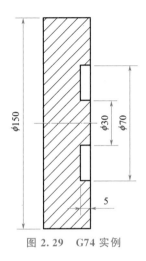

图 2.29　G74 实例

O2430
N10 T0606;　　　　　　　　　（刀具宽度为 4mm）
N20 S1000 M3;
N30 G0 X30. Z2.；
N40 G74 R1.；
N50 G74 X62. Z-5. P3500 Q3000 F0.1；
N60 G0 X200. Z50. M5;
N70 M30;

G 代码编程综合应用实例（图 2.30）：

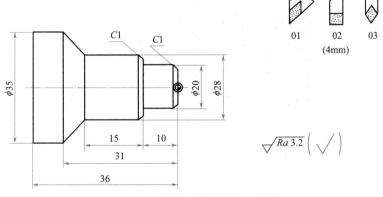

图 2.30　G 代码编程综合应用实例

O2431   (O 指定程序号)
N10 T0101;
N20 S1000 M03;
N30 G00 X40. Z2. ;
N40 G71 U1.5 R1. ;   (U 为吃刀量,R 为退刀量,均为半径值)
N50 G71 P60 Q150 U0.5 W0.2 F0.3;   (P 为轮廓开始段号,Q 为轮廓结束段号,U 为 X 向精加工余量,W 为 Z 向精加工余量)
N60 G01 X18. Z0. ;   (轮廓开始)
N80 X20. Z-1. ;   (倒角 C 1,1×45°)
N90 G01 Z-10. ;   (ϕ 20mm)
N100 X26. ;
N110 X28. Z-11. ;   (倒角 C 1,1×45°)
N120 Z-25. ;
N140 X35. Z-31. ;
N150 Z-36. ;   (轮廓结束)
N160 G70 P60 Q150 S1500 F0.05;   (根据轮廓段组 N60～N150 精加工轮廓)
N170 G00 X50. Z60. ;
N180 T0202;
N190 S700 M03;   (切断时需要降速)
N210 G00 X37. Z-40. ;   (准备切削,设对刀点在割刀的左刀尖,则右刃将车出 36mm 长的工件)
N220 G01 X0. F0.03;
N230 G00 X50. ;
N240 G00 Z0. ;
N250 M05;
N260 M02;   (程序结束)

2.4.3 铣削加工中心 MAHO 600C 编程

铣削加工中心 MAHO 600C 能够实现四轴联动(旋转工作台 B 轴与 X、Y、Z 轴组成四坐标机床)。MAHO 600C 的操作面板如图 2.31 所示。MAHO 600C 的 G 代码及其功能见表 2-5。

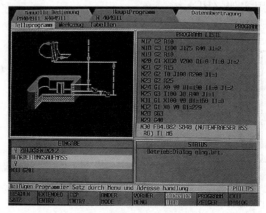

图 2.31　MAHO 600C 的操作面板

表 2-5 MAHO 600C 的 G 代码及其功能

G 代码	功能
G00*	点定位（快速运动）
G01	直线插补
G02	圆弧插补（顺时针）
G03	圆弧插补（逆时针）
G04**	暂停，0s 或 1~983s
G17*	XOY 工作平面定义
G18	XOZ 工作平面定义
G19	YOZ 工作平面定义
G22**	子程序调用
G23**	返回主程序
G25	手动进给倍率控制：开
G26	手动进给倍率控制：关
G27	圆周进给
G28	准确停止进给
G40*	取消刀具半径补偿
G41	刀具半径左补偿
G42	刀具半径右补偿
G43	刀具长度正补偿
G44	刀具长度负补偿
G51*	取消工件坐标系
G52	设置工件坐标系零点（轴复位）
G53*	取消工件坐标系
G54	设置工件坐标系零点 1
G55	设置工件坐标系零点 2
G56	设置工件坐标系零点 3
G57	设置工件坐标系零点 4
G58	设置工件坐标系零点 5
G59	设置工件坐标系零点 6
G63*	取消 G64 激活功能
G64	激活几何图形处理器
G70	英制输入（英寸）
G71*	公制输入（毫米）
G72*	取消镜像/比例放缩
G73	镜像/比例放缩指令（比例定义 A4＝）
G74**	返回参考点
G77**	均布孔指令
G78**	点位定义指令
G79**	循环调用
G81	钻削循环
G83	深孔钻削循环
G84	攻螺纹循环

续表

G 代码	功能
G85	铰削循环
G86	镗削循环
G87	方腔铣削循环
G88	键槽铣削循环
G89	圆腔铣削循环
G90*	绝对坐标编程
G91	增量坐标编程
G92	增量坐标系变换指令（坐标系旋转定义 B4＝）
G93	绝对坐标系变换指令（坐标系旋转定义 B4＝）
G94*	进给速度 mm/min（G71）或 in/min（G70）
G95	回转进给速度 mm/r（G71）或 in/r（G70）
G98	图形窗口定义指令
G99	图形毛坯定义指令

注：1. * CNC 系统在默认读取状态有效。
  2. ** 在程序执行中有效。

铣削编程实例 1（图 2.32）：

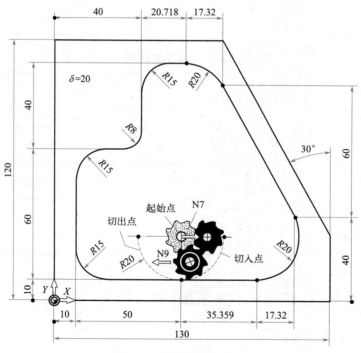

图 2.32 铣削编程实例 1

```
% PM24301
N24301
N1 G17 S900 T31 M66;          （铣刀直径为 φ10mm，手动换刀）
N2 G54;                        （工件坐标零点）
```

```
N3  G98 X-10 Y-10 Z-20 I150 J140 K30;   (模拟窗口)
N4  G99 X0 Y0 Z-20 I130 J120 K20;       (毛坯大小)
N5  G00 X60 Y30 Z8 M3;
N6  G01 Z-21 F50;                       (进刀到给定深度)
N7  G43 X80 F100;                       (刀具长度正 X80)
N8  G42;                                (刀具半径右补偿)
N9  G02 X60 Y10 R20;                    (或 I60 J30)
N10 G01 X25;
N11 G02 X10 Y25 R15;                    (或 I25 J25)
N12 G01 Y55;
N13 G02 X25 Y70 R15;                    (或 I25 J55)
N14 G01 X32;
N15 G03 X40 Y78 R8;                     (或 I32 J78)
N16 G01 Y95;
N17 G02 X55 Y110 R15;                   (或 I55 J95)
N18 G01 X60.718;
N19 G02 X78.039 Y100 R20;               (或 I60.718 J90)
N20 G01 X112.679 Y40;
N21 G02 X95.359 Y10 R20;                (或 I90.359 J30)
N22 G01 X60;
N23 G02 X40 Y30 R20;                    (或 I60 J30)
N24 G40;                                (取消刀具半径补偿)
N25 G00 Z50 M5;                         (退刀位置,主轴停止)
N26 G53;                                (取消 G54)
N27 T0 M66;                             (手动取下刀具)
N28 M30;
```

1. G40/G41/G42：刀具补偿

在数控加工中，如果刀具中心沿工件轮廓编程路径移动，由于没有考虑刀具直径，因此加工出来的工件尺寸不正确。

现代数控系统都有刀具半径补偿功能。系统需要编程路径、刀具直径和刀具相对于轮廓的位置，设定刀具偏置参数。因此，按被加工零件轮廓尺寸进行程序编写，不必考虑刀具直径，在对刀时需要将刀具偏置参数输入数控系统。

G40 取消刀具补偿。

G41 刀具半径左补偿：沿刀具运动方向看，刀具在工件加工轮廓左侧。

G42 刀具半径右补偿：沿刀具运动方向看，刀具在工件加工轮廓右侧。

2. G14：程序段重复指令

指令格式：G14 N1= __ N2= __ J __

其中，N1——重复执行的起始程序号；

N2——重复执行的终止程序号；

J——重复执行的次数（缺省为重复 1 次）。

3. G92/G93：坐标系变换指令（图 2.33）

指令格式：G92 X＿ Y＿ Z＿ B1=＿ L1=＿ B4=＿

指令格式：G93 X＿ Y＿ Z＿ B2=＿ L2=＿ B4=＿

其中，X，Y，Z——直线坐标的平移量；

  B1，L1——极坐标（G92）；

  B2，L2——极坐标（G93）；

  B4——旋转后坐标轴与旋转前的坐标轴的夹角。

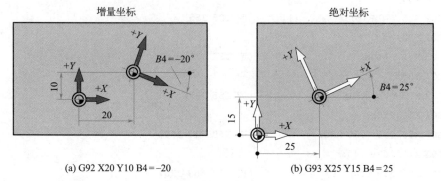

图 2.33　G92/G93 功能

铣削编程实例 2（图 2.34）：

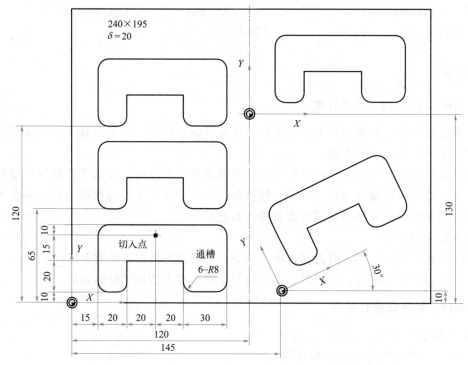

图 2.34　铣削编程实例 2

```
% PM24302
N24302
N1 G17 S900 T31 M66;              (铣刀直径为φ10mm)
N2 G54;
N3 G98 X-10 Y-10 Z-10 I260 J215 K30;
N4 G99 X0 Y0 Z-20 I240 J195 K20;   (毛坯定义)
N5 G00 X55 Y45 Z2 M13;
N6 G1 Z-21 F50;                    (进刀)
N7 G43 Y55 F100;                   (刀具长度正补偿Y55)
N8 G42;                            (刀具半径右补偿)
N9 G01 X97;                        (加工轮廓)
N10 G02 X105 Y47 R8;
N11 G01 Y18;
N12 G02 X97 Y10 R8;
N13 G01 X83;
N14 G02 X75 Y18 R8;
N15 G01 Y30;
N16 G01 X35;
N17 G01 Y18;
N18 G02 X27 Y10 R8;
N19 G01 X23;
N20 G02 X15 Y18 R8;
N21 G01 Y47;
N22 G02 X23 Y55 R8;
N23 G01 X55;                       (轮廓加工结束)
N24 G00 Z50;
N25 G40;                           (取消刀具半径补偿)
N26 G92 Y55;                       (增量坐标系零点变换)
N27 G14 N1= 5 N2= 26 J2;           (重复循环两次)
N28 G93 X120 Y130;                 (绝对坐标系变换)
N29 G14 N1= 5 N2= 25;              (重复循环一次)
N30 G93 X145 Y10 B4= 30;           (绝对坐标系变换)
N31 G14 N1= 5 N2= 25;              (重复循环一次)
N32 G00 Z50 M5;                    (退刀)
N33 G53;                           (取消G54)
N34 M30;
```

4. G72/G73：镜像/比例放缩指令

G72：取消镜像/比例放缩。

G73：镜像，坐标轴号后跟"-1"[G73 X-1(Y-1)(Z-1)]，表示相应的坐标为相反值，如图2.35所示；

比例放缩（如G73 A4＝__，放缩比例因子A4＝）。

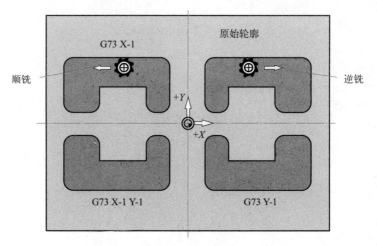

图 2.35 G73 功能

G73 镜像实例（图 2.36）：

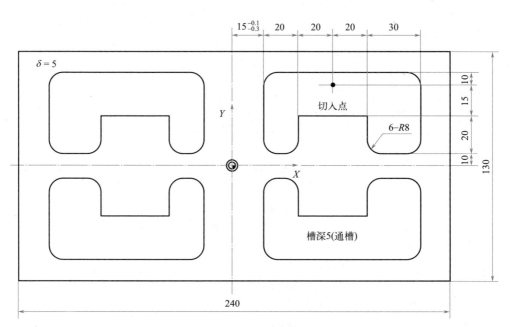

图 2.36 G73 镜像实例

```
% PM24303
N24303                              (镜像实例,毛坯:240mm×130mm×5mm)
N1 G17 S900 T31 M66;                (铣刀直径为 φ10mm)
N2 G54;
N3 G98 X-130 Y-75 Z-10 I260 J150 K20;
```

```
N4 G99 X-120 Y-65 Z-5 I240 J130 K5;        (毛坯定义)
N5 G0 X55 Y45 Z2 M13;                       (G00 可以简写为 G0)
N6 G1 Z-6 F50;                              (进刀)
N7 G43 Y55 F100;                            (刀具长度正补偿 Y55)
N8 G42;                                     (刀具半径右补偿)
N9 G1 X97;                                  (加工轮廓)
N10 G2 X105 Y47 R8;
N11 G1 Y18;
N12 G2 X97 Y10 R8;
N13 G1 X83;
N14 G2 X75 Y18 R8;
N15 G1 Y30;
N16 G1 X35;
N17 G1 Y18;
N18 G2 X27 Y10 R8;
N19 G1 X22.8;                               (尺寸 15 取中间公差)
N20 G2 X14.8 Y18 R8;
N21 G1 Y47;
N22 G2 X22.8 Y55 R8;
N23 G1 X55;                                 (轮廓加工结束)
N24 G0 Z50;
N25 G40;                                    (取消刀具补偿)
N26 G73 X-1;                                (对 Y 轴作镜像)
N27 G14 N1=5 N2=25;                         (重复一次)
N28 G72;
N29 G73 X-1 Y-1;                            (对原点作中心对称)
N30 G14 N1=5 N2=25;                         (重复一次)
N31 G72;
N32 G73 Y-1;                                (对 X 轴作镜像)
N33 G14 N1=5 N2=25;                         (重复一次)
N34 G72;                                    (取消镜像)
N35 G0 Z50 M5;                              (退刀)
N36 G53;
N37 M30;
```

2.4.4 FANUC 0MC 数控铣削编程

FANUC 0MC 数控系统 G 代码及其功能见表 2-6。

表 2-6　FANUC 0MC 数控系统 G 代码及其功能

G 代码	功能	G 代码	功能
G00	快速定位	G51	比例放缩
G01	直线插补	G52	局部坐标系设定
G02	顺时针圆弧插补（CW）/顺时针螺旋插补	G53	机床坐标系选择
G03	逆时针圆弧插补（CCW）/逆时针螺旋插补	G54	工件坐标系选择 1
G04	暂停	G55	工件坐标系选择 2
G09	精确停止	G56	工件坐标系选择 3
G10	可编程数据输入（资料设定模式）	G57	工件坐标系选择 4
G11	取消可编程数据输入（资料设定模式取消）	G58	工件坐标系选择 5
G15	极坐标指令取消	G59	工件坐标系选择 6
G16	极坐标指令	G65	宏程序调用
G17	XOY 工作平面定义	G66	模态宏程序调用
G18	XOZ 工作平面定义	G67	模态宏程序调用取消
G19	YOZ 工作平面定义	G68	坐标系旋转/三维坐标系变换
G20	英制输入（英寸）	G69	坐标系旋转取消/三维坐标系变换取消
G21	公制输入（毫米）	G73	高速深孔钻孔循环
G22	行程检查功能打开（ON）	G74	左旋螺纹攻螺纹循环
G23	行程检查功能关闭（OFF）	G76	精镗孔循环
G24	镜像功能	G80	固定循环取消/外部操作功能取消
G25	镜像功能取消	G81	钻孔循环（钻中心孔）
G27	参考点返回检查	G82	钻孔循环（带停顿）
G28	返回参考点	G83	深孔钻孔循环
G29	从参考点自动复位	G84	攻螺纹循环
G30	返回第二、第三、第四参考点	G85	粗镗孔循环
G31	跳转功能	G86	镗孔循环
G33	螺纹切削	G87	反镗孔循环
G39	转角补正圆弧切削	G90	绝对坐标编程
G40	刀具半径补偿取消/三维补偿取消	G91	增量坐标编程
G41	刀具半径左补偿/三维补偿	G92	工件坐标系设定/主轴最大转速设定
G42	刀具半径右补偿	G94	进给速度：每分钟进给量（mm/min）
G43	刀具长度正补偿	G95	进给速度：每转进给量（mm/r）
G44	刀具长度负补偿	G96	恒线速度控制（mm/min）
G49	刀具长度补偿取消	G97	恒线速度控制取消（r/min）
G50	取消比例放缩	G98	固定循环返回起点

1. G00：点定位（快速运动）

指令格式：G00 X __ Y __ Z __

2. G01：直线插补

指令格式：G01 X __ Y __ Z __ F __

3. G02/G03：顺时针/逆时针圆弧插补或螺旋插补

指令格式：G02 X __ Y __ R __ F __（如果 G17:X __ Y __,G18:X __ Z __,G19:Y __ Z __）
　　　　　 G02 X __ Y __ I __ J __ F __

其中，X __ Y __ Z __ 为圆弧终点坐标，采用 G90 指令时是相对于工件零点的坐标，采用 G91 指令时是相对于圆弧起点的坐标。

I __ J __ K __ 是圆弧的圆心坐标值，分别对应 X __ Y __ Z __。

4. G04：暂停

指令格式：G04 X __ 或 G04 P __（暂停时间或指定速度）

如果指定速度暂停，则需要其他参数，这里不作叙述。

G04 X3.5 或 G04 P3500，表示暂停 3.5s，X 后面可带小数点，单位为 s；P 后面数字不能带小数点，单位为 ms。

5. G40/G41/G42：刀具半径补偿

指令格式：G40 G01 X __ Y __
　　　　　 G41(G42)G01 X __ Y __ D __

其中，D——刀具号，存有预先由 MDI 方式输入的刀具半径补偿值（1~3 位数）。

G41 为刀具半径左补偿指令。

G42 为刀具半径右补偿指令。

G40 为取消刀具半径补偿指令。

采用刀具半径补偿且当刀具移动时，根据存放在数控系统中的刀具半径值，自动计算刀具中心轨迹。

刀具半径补偿时，刀具从起点接近工件，在编程轨迹基础上，刀具中心向左（G41）或向右（G42）始终偏离一个偏置量（刀具半径）的距离，偏置量位于刀具运动轨迹法线方向，从刀具中心轨迹指向编程轨迹。

6. G92：工件坐标系设定/主轴最大转速设定

指令格式：G92 X __ Y __ Z __

将工件坐标系设定在相对于刀具起点的某一空间点上。

G92 工件坐标系设定实例（图 2.37）：

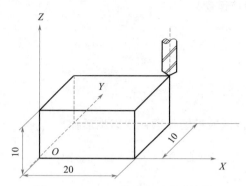

图 2.37　G92 工件坐标系设定实例

G92 X20.Y10.Z10.;

7. G53：机床坐标系选择

指令格式：G53 G90 X __ Y __ Z __

选择机床坐标系后，刀具通过快速运动返回机床零点。G53 使刀具快速定位到机床坐标系中的指定位置上，它是单次指令，仅用于设定机床坐标系的程序段。

G53 实例（图 2.38）：

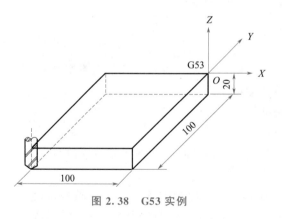

图 2.38　G53 实例

G53 G90 X-100.Y-100.Z-20.;

8. G54/G55/G56/G57/G58/G59：工件坐标系选择 1～6 号

指令格式：G54 G90 G00 (G01)X __ Y __ Z __ (F __)

机床上电并且返回参考点后，这些指令可以分别用来选择 1～6 号相应的工件坐标系。机床开机后默认工件坐标系 1（G54）。

1～6 号工件坐标系通过 CRT/MDI 方式设置。

工件坐标系选择实例（图 2.39）：

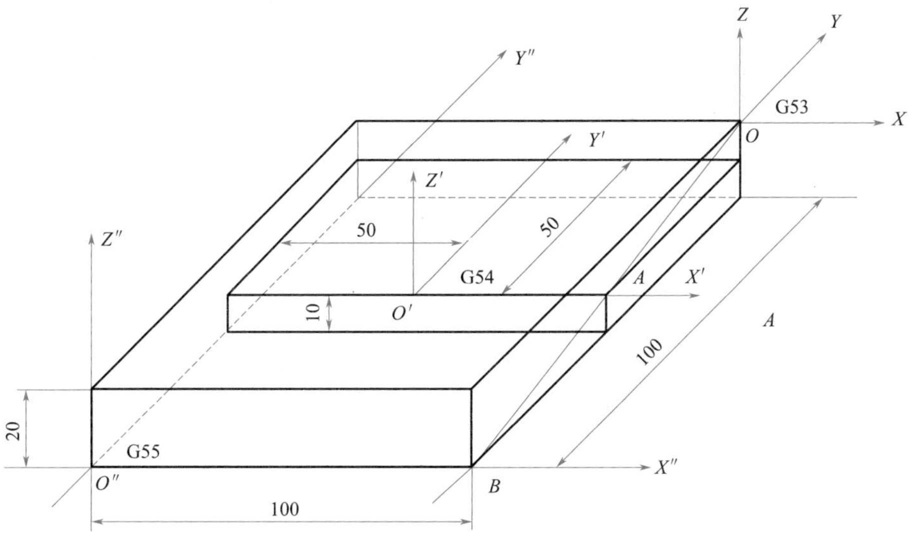

图 2.39 工件坐标系选择实例

G54:X-50.Y-50.Z-10.；
G55:X-100.Y-100.Z-20.；(建立原点在O'的G54工件坐标系和原点在O"的G55工件坐标系)
N10 G53 G90 X0.Y0.Z0.；
N20 G54 G90 G01 X50.Y0.Z0.F100；
N30 G55 G90 G01 X100.Y0.Z0.F100；(刀尖点的运动轨迹如图2.39中OAB所示)

G55 实例（图 2.40）：

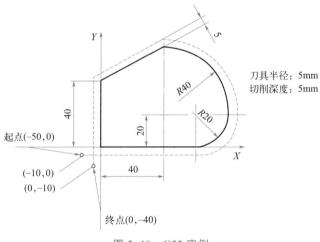

图 2.40 G55 实例

O24401
N10 G55 G90 G01 Z40.F200;   （工件坐标系 2）
N20 M03 S1000;
N30 G01 X-50.Y0.;   （到达X、Y坐标起始点）
N40 G01 Z-5.F100;   （到达Z坐标起始点）

```
N50 G01 G42 X-10. Y0. H01;        （刀具半径右补偿）
N60 G01 X60. Y0. ;                （切入轮廓）
N70 G03 X80. Y20. R20. ;
N80 G03 X40. Y60. R40. ;
N90 G01 X0. Y40. ;
N100 G01 X0. Y-10.                （切出轮廓）
N110 G01 G40 X0. Y-40. ;          （取消刀具半径补偿）
N120 G01 Z40. F200;
N130 M05;
N140 M30;                         （程序结束）
```

9. G68/G69：坐标系旋转（指定旋转中心或旋转角度）

指令格式：G68 X __ Y __ R __

其中，X，Y——旋转中心的坐标值 X、Y、Z，具体是哪个平面由工作平面定义指令确定（G17、G18 或 G19），当 X、Y 省略时，当前的位置即为旋转中心；

　　R——旋转角度，逆时针旋转定义为正方向，顺时针旋转定义为负方向。

G69 为取消坐标系旋转。

数控编程时，需要编程图形按指定旋转中心及旋转方向旋转一定的角度，G68 指令实现了该功能。例如，工件需要加工的位置与机床有一夹角，可以通过坐标系旋转方便编程。另外，如果工件的形状由许多相同的图形组成，则可将图形单元编成子程序，然后用主程序的旋转指令调用。

G68 实例（图 2.41）：

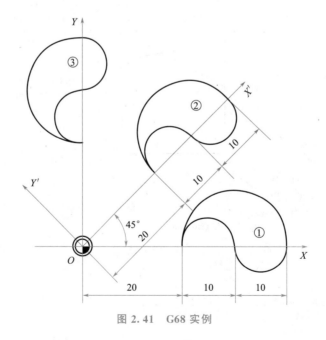

图 2.41　G68 实例

```
O24402                    (主程序)
N10 G90 G17 S1000 M03;
N20 M98 P1100;            (①)
N30 G68 X0. Y0. P45;      (旋转 45°)
N40 M98 P1100;            (②)
N50 G69;                  (取消旋转)
N60 G68 X0. Y0. P90;      (旋转 90°)
M70 M98 P1100;            (③)
N80 G69 M05 M30;          (取消旋转)

子程序
O1100
N100 G90 G01 X20. Y0. F100;
N110 G02 X30. Y0. R5. ;
N120 G03 X40. Y0. R5. ;
N130 X20. Y0. R10. ;
N140 G00 X0. Y0. ;
N150 M99;
```

10. G90/G91：绝对坐标编程指令/增量坐标编程指令

刀具运动轨迹坐标定义有两种：绝对坐标和增量坐标。在绝对坐标下，程序段中的尺寸为绝对坐标值；在增量坐标下，程序段中的尺寸为增量坐标值，即相对于前一工作点的增量值。

11. M98/M99：子程序调用/结束

指令格式：M98 P＿＿

其中，P——子程序调用情况。P后共有8位数字，前4位为调用次数，省略时为调用1次；后4位为所调用的子程序号。

M99 表示子程序结束，并返回调用子程序的主程序中。

12. G51/G50：比例缩放/比例缩放取消

指令格式：G51 X＿＿ Y＿＿ Z＿＿ P＿＿

其中，X，Y，Z——比例中心坐标（绝对坐标方式）；

　　　　P——比例系数，最小输入量为 0.001，比例系数为 0.001～999.999；该指令以后的移动指令，从比例中心点开始，实际移动量为原数值的 P 倍；P 值对偏移量无影响。

13. G24/G25：镜像/镜像取消

G24 实例（图 2.42）：

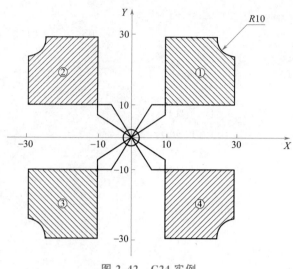

图 2.42 G24 实例

O24403(主程序)

N10 G91 G17 S1000 M03;

N20 M98 P1101;　　　　　(①)

N30 G24 X0.;　　　　　　(Y 轴镜像,X0)

N40 M98 P1101;　　　　　(②)

N50 G24 X0.Y0.;　　　　 (X 轴、Y 轴镜像,镜像位置为原点)

N60 M98 P1101;　　　　　(③)

N70 G25 X0.;　　　　　　(取消 Y 轴镜像)

N80 G24 Y0.;　　　　　　(X 轴镜像,Y0)

N90 M98 P1101;　　　　　(④)

N100 G25 Y0.;　　　　　 (取消镜像)

N110 M05;

N120 M30;

子程序

%1101

N200 G41 G00 X10.0 Y4.0 D01;

N210 Y1.0

N220 Z-98.0;

N230 G01 Z-7.0 F100;

N240 Y25.0;

N250 X10.0;

N260 G03 X10.0 Y-10.0 I10.0;

N270 G01 Y-10.0;

N280 X-25.0;

N290 G00 Z105.0;

N300 G40 X-5.0 Y-10.0;

N310 M99;

2.5 计算机辅助制造

在手工编程中，数控程序员通过数控代码确定机床或刀具的运动，如果是三维图，那么手工编程将很困难。

多年来，国内外一直在研究自动编程，随着 CAD/CAM 技术的发展，交互式图形编程集成到许多数控系统中。交互式图形编程通过 CAM 系统仿真模块或 CAD/CAM 数据库建立几何模型，其菜单驱动技术提高了数控编程用户界面的友好性。刀具运动控制指令实现数控加工时刀位的自动计算。数控程序员能够通过 CAM 系统仿真模块验证刀具轨迹，这大大提高了刀具轨迹的计算精度和效率。

CNC 机床加工零件的流程如图 2.43 所示。

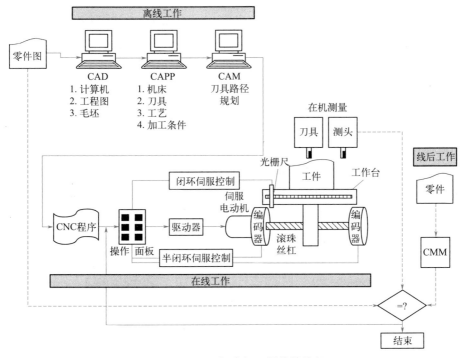

图 2.43 CNC 机床加工零件的流程

整个流程可以分解为如下三种类型。

(1) 离线工作：CAD、CAPP、CAM。
(2) 在线工作：数控加工、监控和在机测量。
(3) 线后工作：CAI 后操作。

离线工作用于生成控制数控加工程序。在离线阶段，确定零件图后，通过二维或三维 CAD 创建该零件的几何模型。这里的 CAD 是指 CAD 建模，包括优化设计和结构分析，这些工作不能在车间现场完成。

完成 CAD 建模后，对生成的加工信息进行工艺规划。在此阶段，选择机床、刀具、夹具及固定装置等，确定加工参数和加工顺序，因为工艺设计难度高，CAPP 在技术上还不成熟，工艺设计主要依赖于工艺人员的经验和技术水平。

CAM 是生成数控程序的最后阶段。在此阶段，根据 CAD 的几何信息和 CAPP 的加工信息生成刀具路径。在刀具轨迹生成过程中，要进行刀具与工件的干涉检查、减少加工和换刀时间、考虑机床性能的影响等。对于三轴以上的数控机床，CAM 提供 2.5D 编程和 3D 编程。

在线工作是指数控机床完成零件加工的流程。在离线阶段，已经生成数控程序，用户仅能在 CNC 系统中进行简单的程序编制，CNC 系统从存储器读取并解释（译码）数控程序，控制各坐标轴运动。CNC 系统根据数控程序生成位置和速度控制指令，按指令控制伺服电动机运动。当伺服电动机的转动经滚珠丝杠机构转化为直线运动后，带动工件或刀具移动，实现零件的加工。

为了提高加工精度，不仅要保证伺服电动机、工作台导轨、滚珠丝杠和主轴的精度，还要保证机床结构的刚性。设计机械结构时要考虑振动和温度的影响。CNC 系统中的编码器和传感器的性能及控制机构也对加工精度有影响。

在线工作包括加工过程和机床状态的监控。在实际加工时，刀具磨损检测，热变形补偿，自适应控制，基于切削力、热、电流监测的刀具变形补偿等都在该阶段完成。在机测量功能用于加工误差控制，通过检查加工完的零件，将数据反馈给 CNC 系统进行误差补偿。

线后工作主要是通过 CAI 对成品进行检验。在此阶段，常使用 CMM（coordinate measurement machine，三坐标测量机）进行测量，将结果与几何模型进行比较，实现误差修正。误差修正工作通过修改刀具补偿值或后续工序（如重新加工和磨削）来完成。这一阶段，测量零件的形状并根据测量数据生成几何模型的过程是一个逆向工程。综上所述，通过以上阶段的工作，CNC 机床完成高精度和高效生产，不仅可以加工形状简单的零件，而且可以加工形状复杂的零件。CNC 机床在加工零件时，可以通过修改程序以适应不同零件的加工，通过存储的程序实现重复加工，所以，与普通机床相比，CNC 机床的适应性更强。

广泛应用的 CAM 系统很多，其工作流程图如图 2.44 所示。

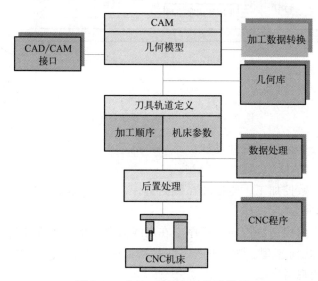

图 2.44　CAM 系统的工作流程图

CAM 系统的基本特征如下。

(1) 几何模型/CAD 接口。

工件的几何形状可以由基本的几何要素（如点、线、弧、样条曲线或曲面）定义。二维或三维几何要素以数学模型的形式存储在计算机中。数学模型可以是线框模型、曲面模型，也可以是实体模型。

几何模型可以通过标准的 CAD/CAM 接口格式（如初始化图形交换规范）从其他 CAD/CAM 系统中导入数据。初始化图形交换规范是美国国家标准与技术研究院和工业界在空军的支持下共同编制的图形交换标准，它提供了三维软件系统的不同文件转换格式。

在整个 CAD/CAM 系统中，几何要素可以转换为标准（中性）文件，然后由该标准文件转换为其他系统可读取的格式的文件。

(2) 刀具轨迹定义。

几何模型完成后，将工件的作业设置、操作设置、运动定义等加工数据输入计算机，生成加工工件的刀位点文件（CL 文件）。

① 作业设置：为 CL 文件输入机床基准、初始位置和刀具直径。

② 操作设置：将进给速度、几何公差、刀具接近/退出平面、主轴转速、冷却液开/关、刀具偏置和刀具选择等操作参数输入系统。

③ 运动定义：内置加工命令用于控制刀具运动以完成切削加工，包括加工安排、轮廓加工、槽加工、曲面加工、过切检查等。

(3) 数据处理。

输入数据被转换为数控加工数据。计算机处理曲面、刀具偏置等几何信息和工艺信息，计算加工走刀过程中的刀位点（CL 点），从而生成 CL 文件。刀具路径通常需要调用仿真模块模拟刀具切削过程进行验证。

此外，要完成刀具清单、装配单、工时定额等生产计划数据，以供用户参考。

(4) 后置处理。

不同 CNC 机床的功能差别很大，编程格式和指令也千差万别。需要将 CAM 编程生成的通用 CL 文件转换为特定机床能够识别的 CNC 程序，这个过程称为后置处理。后置处理器是结合特定的 CNC 机床把系统生成的刀具轨迹转换为 CNC 机床能够识别的 G 代码的处理器。后置处理器一般分为以下两种。

① 专用后置处理器。专用后置处理器只能生成唯一指定 CNC 机床的数控加工程序，用户无权修改。

② 通用后置处理器。通用后置处理器的后置处理程序语法规则是通用的，可以由用户针对指定 CNC 机床进行二次开发。

(5) 数据传输。

后置处理后，通过离线或在线方式，将数控程序下载到 CNC 机床中。

① 离线方式。通过数据媒介将数控程序传输到 CNC 机床，辅助存储设备包括穿孔带、磁带或磁盘。

② 在线方式。在线方式常用于 DNC，数据传输通过数据传输电缆，以串行通信或并行通信方式完成。

异步串行通信是一种主要的计算机通信方式，其中 RS-232C 标准应用最为广泛。RS-232C 接口（9 针或 25 针）是计算机的标配，RS-232C 接口价格低、编程容易，它的波特率最高可达 38400，但它的噪声容限只有 15m。

并行通信通常用于计算机和外部设备之间的通信，如传感器、可编程控制器（PLC）和执行器。最常见的标准是 IEEE488，它采用 24 针连接器，包括 8 条数据线、8 条控制线和 8 条地线，最大数据传输速率为 1Mbit/s，最大传输距离为 20m。

为了使 CAD/CAM 系统能够顺利运行，需要将这些设备连接在一起。在 LAN 中，终端可以访问网络上的任何计算机或车间设备，而不需要物理硬件，速度可达 300Mbit/s。例如，以太网以 100Mbit/s 的速度运行，比 RS-232 串行通信（115.2kbit/s）快得多。

LAN 包括网络软件和硬件，由一组称为网络协议的规则控制。软件能使硬件产生和接收信号，网络传输介质携带信号的同时，能够控制数据的处理和故障的恢复。网络协议定义了网络的逻辑、电气和物理连接特性。为了在网络中实现有效的通信，必须遵循相同的网络协议。

本 章 小 结

本章介绍了数控加工工艺特点、数控加工的基础知识、数控编程的内容和步骤、车削加工中心 MAHO GR350C、FANUC 0i MATE TB 数控车削编程、铣削加工中心 MAHO 600C、FANUC 0MC 数控铣削编程。

通过本章学习，要求学生了解数控编程的工艺处理、数值计算等基本方法，初步掌握数控加工工艺设计及切削用量选择，掌握常用 G 代码，能够完成数控机床加工、调试等操作，能够拓展 CAM 编程，掌握数控技术常用词汇，如 CNC programming、coordinate system、absolute coordinate system、incremental coordinate system、machine zero point、reference point、G-code、tool compensation、CAM 等。

本章的知识和能力图谱如图 2.45 所示。

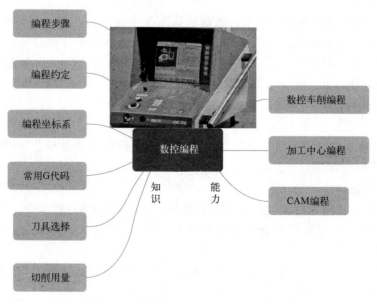

图 2.45 数控编程的知识和能力图谱

第 3 章

CNC 装置和控制原理

教学目的及要求

了解 CNC 装置的硬件结构；

了解 CNC 系统控制软件；

了解插补原理；

掌握刀具补偿原理；

了解 CNC 系统的加减速控制；

了解 CNC 系统中的 PLC。

3.1 CNC 装置的硬件结构

CNC 装置是 CNC 系统的核心。CNC 装置的硬件结构与普通计算机系统的硬件结构类似，主要由微处理器、存储模块、I/O 接口和位置控制模块等组成。CNC 装置通过控制以下设备操作 CNC 机床：MCU、位置传感器、刀具夹持结构、工件夹持机构和机床本体，如图 3.1 所示。

CNC 系统各模块间的数据流如图 3.2 所示。译码器与粗插补器、粗插补器与加减速控制器、加减速控制器与精插补器之间的数据均通过缓冲寄存区传递。每个缓冲寄存区包含的数据参见图 3.2。精插补器和位置控制器使用全局变量实现数据传递。

CNC 系统的特点如下。

(1) 存储多个数控程序。随着计算机存储技术的发展，新的 CNC 系统有足够的容量存储多个程序。CNC 系统的生产厂家为 MCU 提供一个或多个内存扩展接口。

(2) 程序输入方式多样化。传统的硬线 MCU 仅限于穿孔带作为输入控制介质，而 CNC 系统提供多种程序输入方式，如穿孔带、磁带、软盘、与计算机的 RS-232 串行通信和手动数据输入（操作面板输入）。

(3) 在机程序编辑。CNC 程序保存在 MCU 计算机内存中，在数控加工现场可以调

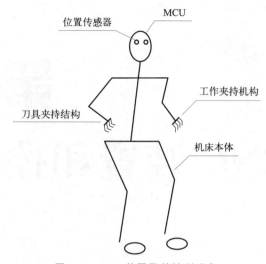

图 3.1　CNC 装置及其控制设备

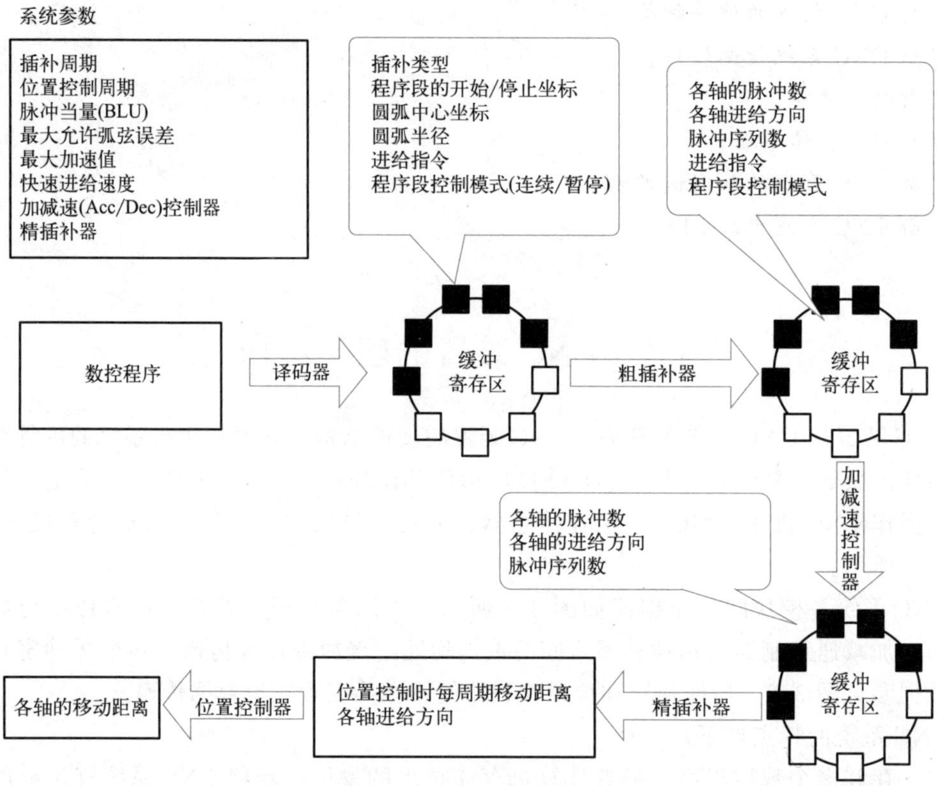

图 3.2　CNC 系统各模块间的数据流

出来进行编辑。CNC 程序可以在数控加工现场进行调试和修改，不用返回编程室。除了能够修改 CNC 程序外，编辑模式下还可以优化加工过程的切削参数。CNC 程序经过修改和优化后，可以将修改后的版本存储在穿孔带或其他输入控制介质上，以备将来使用。

(4) 固定循环和子程序编程。内存容量的提高和控制计算机的编程能力有利于将常用加工循环存储为宏程序，方便数控程序调用。数控程序员不必将需要重复的一组固定加工动作指令写入程序中，只需用 call 语句调用，并在某个程序段执行宏程序（循环加工）。这些固定循环通常要求定义某些参数，如螺孔加工必须指定螺孔的直径、螺孔的间距和其他参数。

(5) 插补。由于插补运算的实时性要求，插补算法通常由 CNC 系统完成。直线插补和圆弧插补可通过硬件固化到控制装置，但螺旋插补、抛物线插补和三次样条插补通常由算法程序执行。

(6) 设置定位功能。加工中心与夹具的作用不仅是夹紧工件，而且要以各方向的定位面作为参考基准，确定工件的编程原点。安装夹具时，要求工件对加工中心各坐标轴线上的角度进行准确调整。

使用 CNC 系统中的软件进行定位可以简化夹具安装，工件的定位是 CNC 机床的优势之一，工件位置确定了，操作者无须以极高的精度将夹具安装在机床工作台上。相反，机床坐标系需要通过夹具的坐标位置或工件的定位来建立。

(7) 刀具长度和尺寸补偿。在传统数控系统中，必须精确设置刀具尺寸，以符合数控程序中刀具轨迹的定义。现在利用 CNC 系统进行刀具轨迹精确控制的方法有两种：一种是手动将刀具尺寸输入 MCU，这些值可能与最初编程的设定尺寸不同，需要在刀具轨迹计算中自动进行补偿；另一种是在机床中使用刀具长度传感器，刀具安装在主轴上，传感器自动测量其长度，该测量值随后用于校正编程的刀具轨迹。

(8) 加速和减速计算。当机床高速进给加工时，此功能适用。它的作用是避免刀具轨迹突然变化时由于机床惯性而在加工面上产生刀痕。因此，在刀具轨迹变化时进给速度要平滑减小，然后在方向变化后加速回到编程进给速度。

(9) 串行通信接口。随着信息的网络化和可视化成为一种必然趋势，大多数 CNC 系统都配备标准的 RS-232 串行通信接口或其他通信接口，将机床与其他计算机或驱动设备连接起来。这提高了程序的通用性，如从信息中心下载数控程序、汇总工件数量、加工次数和机床利用率等、与外部设备（装卸机器人）通信。

(10) 故障诊断。许多 CNC 系统具有故障诊断功能，监测机床的各模块，进行故障检测、故障预防和故障诊断。

3.1.1 CNC 系统中的 MCU

MCU 是 CNC 系统与传统数控系统的主要区别。CNC 系统中 MCU 的结构如图 3.3 所示。

MCU 由以下子系统组成。

(1) CPU。

(2) 内存。

(3) I/O 接口。

(4) 机床坐标轴的位置控制和主轴转速控制。

(5) 机床顺序控制。

MCU 的各子系统通过系统总线连接在一起，系统总线在网络组件之间传输数据和信

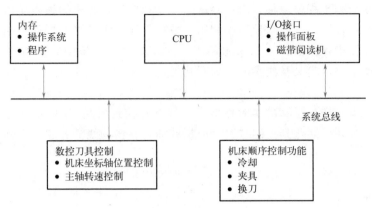

图 3.3 CNC 系统中 MCU 的结构

号。MCU 是 CNC 系统的核心，它有两个子单元：数据处理单元（data processing unit，DPU）和循环控制单元（control loop unit，CLU）。

1. CPU

CPU 是 MCU 的"大脑"。它基于主存储器中的程序来管理 MCU 中的其他组件。CPU 可分为三个部分：控制单元、算术逻辑单元（arithmetic logic unit，ALU）、随机存储器（RAM）。

控制单元从存储器中检索命令和数据进行分析，并且发出相应的 MCU 控制信号，简而言之，它对 MCU 的所有活动进行排序、协调和调节。ALU 由执行各种运算（加法、减法和乘法）、计数和存储器中软件所需逻辑运算等电路组成。RAM 是与 CPU 直接交换数据的临时存储器，它与系统总线的主存储器相连。

2. 内存

在 CNC 系统中，CPU 中的 RAM 不用于长期存储 CNC 控制软件或操作所需的各种程序和数据，CNC 系统通常需要使用其他类型的存储设备来满足更大的存储容量需求。与大多数计算机系统一样，CNC 存储器可以分为两类：主存储器和辅助存储器。主存储器（也称主寄存器）由 RAM 和只读存储器（ROM）组成。操作系统软件和机床接口程序通常存储在 ROM 中，这些程序通常由 MCU 制造厂家设置。数控加工程序存储在 RAM 中，RAM 中的当前程序可以被删除，并在加工修改时被新程序替换。辅助存储器（也称大容量存储器或次级存储器）用于存储大程序和数据文件，这些程序和数据文件根据需要被传输到主存储器。常见的辅助存储器是硬盘和移动存储设备，它们已经取代了传统用于存储数控程序的穿孔带。硬盘是永久安装在 MCU 中的大容量存储设备。CNC 辅助存储器用于存储数控程序、宏程序和其他软件。

3. I/O 接口

I/O 接口是 CNC 系统、其他计算机系统与数控操作对象进行信息交换的纽带。顾名思义，I/O 接口向外部设备发送、接收数据和信号。操作面板是机床操作者与 CNC 系统通信的基本接口，它用于输入与数控程序编辑、MCU 操作模式（如程序控制与手动控制）、速度和进给、切削液泵开/关及类似功能相关的命令。操作面板通常集成字母数字键盘或单设一个键盘。I/O 接口还包括（CRT 或 LED）显示器，用于将数据和信息从 MCU 反馈给机床操作员。显示器用于显示程序执行时的当前状态，并就 CNC 系统中的故障发出警报。

如上所述，CNC 程序可以由机床操作员手动输入并存储在中央计算机网站，通过局域网下载到 CNC 系统。无论企业采用何种方式，I/O 接口中必须包括相应的设备，以允许程序输入 MCU 存储器。

4. 机床坐标轴的位置控制和主轴转速控制

每个机床坐标轴的位置和机床主轴转速控制都需要控制装置来实现。MCU 发出位置控制指令，进行相应的变化，输出位置和速度指令电压给伺服电动机。位置控制系统可以分为开环位置控制系统和闭环位置控制系统两大类，每种情况下都需要不同的硬件来支撑。

根据 CNC 机床的类型，主轴驱动系统完成主运动：工件旋转或刀具旋转。车削是工件旋转，而铣削和钻孔是刀具旋转。主轴转速是大多数 CNC 机床的编程参数。MCU 中的主轴转速控制通常由驱动控制电路和速度反馈传感器组成。具体的主轴驱动硬件取决于主轴驱动的类型。

5. 其他机床顺序控制功能

除了控制工作台位置、进给速度和主轴转速外，还可以在数控程序控制下完成一些辅助功能，这些辅助功能通常包括开/关（二进制）控制、互锁和基于离散型数据控制。为了避免 CPU 过载，有时采用 PLC 来管理这些辅助功能的 I/O 接口。

3.1.2 CNC 控制的新技术

1. 图形处理单元

图形处理单元（graphics processing unit，GPU）有时也称视觉处理单元（visual processing unit，VPU），是一种专门的电子电路。GPU 可通过迅速操作及变更存储器来加速帧缓冲器创建图像，然后输出到显示器。GPU 用于嵌入式系统、手机、PC、工作站和游戏机上图像及图形的相关运算工作。GPU 具有并行结构，比通用 CPU 并行处理大规模数据更高效。在 PC 中，GPU 可以置于视频卡上或嵌入主板，在某些 CPU 中，它们还可以嵌入 CPU 芯片。

1999 年，英伟达公司将 GPU 推广开来，该公司将 GeForce 256 作为世界上第一个 GPU 推向市场，它被呈现为具有集成几何转换和光照处理，三角形设置/剪切和渲染引擎的单芯片处理器，能够每秒处理至少 1000 万个多边形。其竞争对手冶天科技有限公司在 2002 年发布了 Radeon 9700，并提出了"视觉处理单元"这一术语。

现代 GPU 大部分晶体管用于与 3D 计算机图形学相关的计算。它们最初用于加速纹理操作和多边形渲染等内存密集型工作，后来添加单元以加速几何计算，如顶点坐标变换（旋转和平移）的计算。GPU 的最新发展包括支持可编程的着色器，它可以使用 CPU 进行顶点操作控制和纹理操作，用过采样和插值技术来减少混叠，提供更广的色彩空间（色域）。由于这些计算大多涉及矩阵和向量运算，工程师和研究人员越来越多地关注 GPU 在非图形计算中的应用。其中一个例子是比特币的生成，GPU 用于求解哈希函数。

除了 3D 硬件，现在的 GPU 还包括基本的 2D 图像加速和帧缓冲功能（通常采用 VGA 兼容模式）。最近推出的英伟达 RTX 5000 显卡，采用了最新的 AMPERE 架构，并结合了先进的 DLSS（深度学习超级采样）技术。

2. 运动控制器

运动控制器控制某个对象（如电动机）的运行方式。通常运动控制器是用数字控制器实现的，也可以只用模拟元件来实现。

运动控制器的作用是根据 CNC 装置的指令控制直线电动机的高速、高精度运动。它使用 PID 控制实现直线电动机的高速和高加速度控制，利用现代控制理论和直线电动机控制软件生产运动控制器，制造厂家一般都拥有开发运动控制器的工业产权。

除决定机床整体性能的 CNC 装置的发展外，向直线电动机发出和传输尽可能精确的指令对于提高加工精度也很关键。为了精确计算出直线电动机的运动，并将结果传送给 CNC 装置，需要一个运动控制器，运动控制器的开发初衷是针对 CNC 机床运动专用控制器开发，而不是在现有控制器上改进。

3. 多轴运动控制技术

多轴数控机床的多个坐标轴可同时加工，多轴联动需要同时协调运动，多轴运动控制卡如图 3.4 所示。以移动两个关节到达新位置的机器人为例，我们可以想象两个关节同步启停的协调运动。

图 3.4　多轴运动控制卡

运动控制器需要一个负载（移动的物体）、一个原动机（使负载移动）、若干个传感器（能够感知运动并监控原动机）和一个控制器，按指令信息使原动机根据需要移动负载。

运动控制器有如下优点。

（1）位置和速度控制精度高。

（2）运动控制快。

（3）快速响应性好。

(4) 提高生产水平。
(5) 运动平稳性好。
(6) 成本低。
(7) 易于自动化集成。
(8) 运动控制过程的集成性好。
(9) 运动控制开发方便。
(10) 维护简单,易于故障诊断和故障排除。
(11) 运动控制一致性好。
(12) 工作效率高。
(13) 工作环境友好。

3.2　CNC 系统控制软件

CNC 机床由两大部分组成：机床本体和 CNC 装置,后者是机载计算机。这些计算机可能由同一家公司生产,也可能由不同公司生产。通用数字、发那科、通用电气、德马吉、辛辛那提米拉克龙和西门子等都为机床制造商提供 CNC 装置。每个 CNC 装置都有一套标准的固定代码,如启动程序、基本系统代码、加工和测量反馈程序等,其他代码系统可由机床制造厂家添加。因此,程序代码在不同的机床之间有些差异,每台 CNC 机床,无论制造厂家是谁,都由 CNC 装置集成在一起。

在 CNC 系统中,软件种类繁多,主要分为用户软件和控制软件两大类。图 3.5 所示为 CNC 系统控制软件的框架。

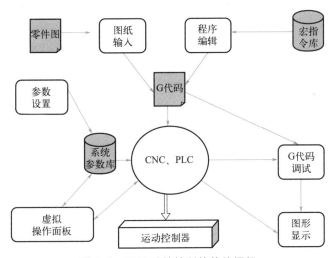

图 3.5　CNC 系统控制软件的框架

1. 用户软件

用户软件又称数控程序,它是用数控语言（如 APT）编写和输入的,用来表示零件的加工过程。G 代码和 M 代码按顺序排列编译成一个数控程序。

2. 控制软件

CNC 系统的主要功能由控制软件实现。控制软件通过相应的子程序完成以下工作。

（1）编译用户输入的程序。

CNC 系统控制软件对使用 ISO 或 EIA 代码的零件数控程序进行编译、整理，并按规定的格式存放；对数控程序的加工指令进行译码，对其坐标数据进行十进制到二进制的转换，对编程轨迹进行刀具半径偏移计算，以得到刀具中心的运动轨迹；对插补计算及速度控制过程中要用到的一些常数进行预计算；等等。不同功能的 CNC 系统，其预处理具体内容不尽相同，但其目的都是相同的。即为插补运算节约时间。CNC 系统对输入数据处理程序的实时性要求不高，它可在加工前或加工过程的空闲时间进行。输入数据处理得越充分，加工过程中实时性较强的插补运算及速度控制越平稳。

（2）插补计算。

插补计算子程序与 CNC 系统中的硬件插补器具有相同的功能，即实现坐标轴脉冲分配功能。插补计算是实时性很强的程序，要尽可能减少该程序中的指令条数，即缩短进行插补运算的时间。因为这个时间直接决定了插补进给的最高速度。在有些 CNC 系统中还采用粗插补与精插补相结合的方法，软件只作粗插补，即每次插补一个小线段，硬件再将小线段分成单个脉冲输出，完成精插补。这样既可提高进给速度，又能有更多的时间进行必要的数据处理。

（3）刀具补偿。

所有类型的 CNC 机床都需要某种形式的刀具补偿，如刀具长度补偿、刀具半径补偿和刀尖半径补偿等。尽管在不同类型的机床上应用形式不同，但所有形式的补偿都允许 CNC 用户针对与刀具相关的不可预测的情况进行修改。一般来说，如果 CNC 用户在编程过程中遇到任何不可预测的情况，则 CNC 制造厂家有义务提出补偿解决方法。

（4）速度控制。

速度控制的目的是控制脉冲分配的速度，即根据给定的速度代码（或其他相应的速度指令）控制插补运算的频率，以保证按预定速度进给。当速度突变时，要进行自动加减速控制，避免速度突变造成伺服系统失调。

速度控制既可以通过软件方法（软件定时器方法）来实现，又可以通过硬件方法来实现，即使用速度指令控制振荡器，通过中断或查询的方式使 CNC 装置进行一次插值计算以保证进给速度。此外，用软件对速度控制数据进行预处理，并与硬件的速度积分器相结合，可以实现高性能的恒定合成速度控制，并大大提高插补进给速度。

（5）位置控制。

位置控制在伺服系统的位置回路中进行。这项工作可以通过软件或硬件来完成。位置控制软件的主要工作是在每个采样周期内将插补计算出的理论位置与实际反馈位置进行比较，用其差值控制进给电动机。在位置控制中，通常还要完成位置回路的增益调整、各坐标方向的螺距误差补偿和反向间隙补偿等，以提高机床的定位精度。

3.3 插补原理

本书所讨论的 CNC 机床都默认有至少一个运动轴。通常，数控程序员常需要对两个或多个坐标轴进行控制实现多轴联动，例如，立铣刀用于加工直面、斜面和圆弧面，除个别运动只涉及一个轴，斜线运动和圆弧运动都包含两个运动轴。

生产 CNC 系统的厂家很多，其中西门子、发那科和艾伦-布拉德利对中国的制造自动化产业影响非常大。中国也有自己的数控品牌，如华中数控、蓝天数控和 JD（北京精雕集团数控）等，用于制造业和教育培训。

在中国广泛使用的三种常见的数控系统是发那科数控系统、西门子数控系统和华中数控系统。

在数控发展的早期阶段，程序需要产生一个斜面或圆弧面，其运动必须被分解为一系列很长且非常小的单轴运动，以尽可能致密地形成所需的形状，这种运动通常需要计算机的参与才能实现。随着一种称为"运动插补"功能的出现，编程常见的复杂运动变得简单得多。在如今的 CNC 系统中，控制斜线运动和圆弧运动已相对容易。

对 CNC 系统来说，插补是一个数学概念，即已知起点和终点，估算若干个中间点的值。利用插补进行轮廓控制时，可以基于少量输入数据来精确估计轮廓编程轨迹。

在直线插补中，两点间的插补沿直线的点群来逼近，并沿此直线控制刀具的运动。如果不是直线，那么可以用逼近的方式用许多小线段去逼近曲线，这样每一段小线段就可以用直线插补了，这种方法有局限性，因为要生成轮廓形状，编程时每一段走刀线段都必须足够小（在精度允许范围内），同时点的数量成倍增加。

1. 逐点比较法直线插补

当机床控制两个坐标轴进行直线联动时，CNC 系统会自动完成轨迹计算，所需的只是坐标的起点和终点，便能自动实时填充起点和终点之间的中间点。实际情况是，CNC 系统用一系列非常小的单轴运动从起点到终点进行拟合，这一系列动作就像爬楼梯，每走一小步都要和终点比较一次，最终的结果是一条拟合的直线，如图 3.6 所示。当两个或多个轴坐标有编程运动时，CNC 系统计算出一系列单轴运动，单步步长决定轴的分辨率，步长越小，轴的分辨率越高。

2. 逐点比较法圆弧插补

逐点比较法圆弧插补逐点计算每步的轨迹方向，并将脉冲发送到控制轴，如图 3.7 所示。

已知圆心坐标和半径的圆弧，在第一象限顺时针方向执行圆弧插补，有

$$F_m = X_m^2 + Y_m^2 - R^2 \tag{3.1}$$

单步的方向是根据 F_m、圆弧方向指令和所在象限来确定的。例如，如果在第一象限以顺时针方向执行圆弧插补，则执行的算法如下。

$F_m > 0$，$-Y$：加工点 (X_m, Y_m) 位于圆弧外侧，沿 $-Y$ 方向走一步。

$F_m = 0$，$+X$：加工点 (X_m, Y_m) 位于圆弧上，原则上可以任意方向走一步，这里约定将其和 $F_m > 0$ 一样，沿 $-Y$ 方向走一步。

$F_m < 0$，$+X$：加工点 (X_m, Y_m) 位于圆弧内侧，沿 $+X$ 方向走一步。

上述算法完成一步后，位置 (X_{m+1}, Y_{m+1}) 将更新，并重复该过程，直到刀具到达指令终点位置 (X_f, Y_f)。

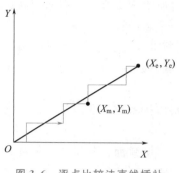

图 3.6　逐点比较法直线插补

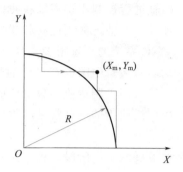

图 3.7　阶梯式逐点比较法插补

3.4　刀具补偿原理

下面讨论不同 CNC 机床的刀具补偿类型。如果作为数控初学者，则有必要了解数控各种形式刀具补偿的目的。事实上，就是要理解为什么需要各种刀具补偿，只有理解了各种刀具补偿类型的目的，才能理解如何使用这些刀具补偿。此外，了解各种刀具补偿的目的后，就可以在任一种刀具补偿中适应其使用流程。

刀具补偿类型及相关机床见表 3-1。

表 3-1　刀具补偿类型及相关机床

刀具补偿类型	相关机床
刀具长度补偿	加工中心
刀具半径补偿	加工中心
夹具偏置	加工中心
刀具偏置量（X、Z 向）	车削中心
刀尖圆弧半径补偿	车削中心
电极丝半径补偿	电火花线切割机
电极丝锥度补偿	电火花线切割机

3.4.1　刀具补偿的定义

编程时，使用刀具补偿能够使编程简单。假设刀具为一个虚拟点，而不必考虑实际使用的刀具长度或刀具半径，直接按零件轮廓数据编程。

CNC 系统可以自动计算铣床的刀具信息（长度和半径）。

1. 铣削和车削的刀具长度补偿

刀具长度补偿相当于设置的零点，用于调整刀具编程长度和刀具实际长度之间的差异。刀具长度是可控的，测量刀具长度 L（刀具安装点 B 和刀尖之间的距离），并将其输入 CNC 装置。

对铣削刀具和车削刀具而言，刀具长度 L 指的都是 Z 轴方向值（图 3.8、图 3.9，R 为刀具半径，Q 为 X 轴方向刀尖到刀具安装点 B 之间的距离）。

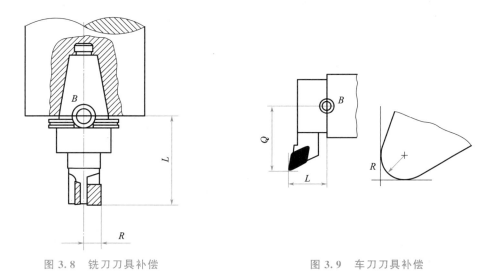

图 3.8　铣刀刀具补偿　　　　　　　　　图 3.9　车刀刀具补偿

在 CNC 系统中，这些刀具补偿存储在存储器中。大多数 CNC 系统可以存储 99 把刀具信息，这些参数在加工过程中通过调用 CNC 程序中的数据来完成激活后才能使用。例如，使用地址 H 或由 T 功能字来指定。

2. 刀具半径补偿

刀具半径补偿（或刀具直径补偿）适用于加工中心和 CNC 机床。采用刀具半径补偿时，数控程序员在编程时不必考虑刀具的半径或直径。与其他刀具补偿一样，刀具半径补偿使编程更简单，因为数控程序员在编写程序时不必考虑刀具的直径。刀具半径补偿还允许在刀具半径变化时不必重新编程，只需修改相应的偏置参数即可（图 3.10）。

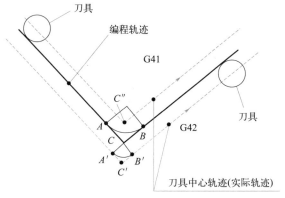

图 3.10　刀具半径补偿

（1）刀具半径补偿的工作原理。

了解 CNC 系统中刀具半径补偿的工作原理是解决各种刀具半径补偿问题的第一步。虽然每个 CNC 制造厂家在处理刀具半径补偿的方式上有些许区别，但本节所述的工作原理适用于大多数 CNC 系统。

刀具半径补偿的三个工作过程如下。

① 建立刀补。

② 刀补（加工）。

③ 撤销刀补。

刀具半径补偿是通过一个命令来设置的，该命令告知 CNC 系统在加工过程中刀具相对于被加工轮廓如何定位。沿刀具运动方向看，刀具在工件加工轮廓的左侧（G41 指令）或右侧（G42 指令）。如果知道顺铣和逆铣的区别，就很容易区分 G41 指令和 G42 指令。如果使用右旋刀具（M03 主轴顺时针旋转），则使用 G41 指令（顺铣）；若使用 G42 指令就是逆铣。一旦设定刀具半径补偿，CNC 系统将使刀具在直线插补（G01）指令和圆弧插补（G02 和 G03）指令下，在轮廓左侧或右侧生成一系列直线和圆弧组合。直线和圆弧组合就是加工出的实际轮廓。

刀具半径补偿直到接收到取消指令一直有效，即刀具保持在工件加工轮廓左侧或右侧加工，直到取消。取消刀具半径补偿的指令是 G40。

（2）刀具半径补偿的局限性。

刀具半径补偿并不适用于所有刀具，它仅适用于能够在刀具周边加工的刀具（如立铣刀、端铣刀、套装铣刀和部分面铣刀），并且仅在加工刀具周边时才需要。钻头、铰刀、丝锥、镗杆和其他轴向加工刀具不需要刀具半径补偿。

在铣削加工过程中，数控程序员需要沿工件的轮廓不断调整刀具方向，轮廓边缘可以是直曲面或仿形曲面等。铣刀是沿工件的轮廓边缘加工的。

数控铣削编程是按铣刀中心轨迹编程的，在这种情况下，数控程序员编程时必须考虑铣刀的直径。在假设刀具没有因为切削力挤压变形而使刀具偏离编程轨迹的情况下，如果铣刀直径为 1in，则编程时刀具中心必须与要铣削的轮廓始终偏移 0.5in 的偏置量。刀具切削力挤压变形机理在这里不作论述。

使用刀具轨迹中心线坐标编程是一种常用的编程方法，但它也有局限性。在刀具半径补偿成为刀具轨迹中心线坐标（工件坐标）之前，这些局限性都会带来实际问题。值得注意的是，刀具轨迹中心线坐标需要在坐标系中对每个坐标轴进行至少一次额外的换算。另外，刀具轨迹中心线坐标通常是从打印图纸中直接获得的。

3.4.2 刀具半径补偿的作用

刀具半径补偿的作用如下。

（1）使编程简单。

（2）粗加工和精加工可用同一个程序。

（3）补偿刀具半径误差。

对于大多数有经验的数控程序员来说，使用直径为 1in 的立铣刀加工被夹在虎钳中的矩形工件的右轮廓是一种简单的加工。如果数控程序员根据刀具轨迹中心坐标编程，而不使用刀具半径补偿，则编程时刀具中心必须与要铣削的轮廓始终偏移 0.5in 的偏置量。

可以想象数控程序员在进行设置时发现找不到 1in 立铣刀，公司只有直径为 0.875in

的立铣刀和直径为 1.25in 的立铣刀。在这种情况下，数控程序员将不得不修改程序。

试切加工时，刀具很少按理想状况加工工件。在加工过程中，刀具、工件、机床本身都承受很大的切削力。加工切削用量越大，切削力就越大，即使铣刀保持相当高的刚性（稳固的立铣刀夹紧和短的总长度），在加工过程中刀具也会有一些偏转，这是因为刀具的切削刃会因为切削力有远离被加工表面的倾向。

如果你曾经刮过房间墙壁上的油漆，你就会理解这种偏转。刮掉油漆时，你要尽最大努力推动刮板以清除油漆，但很多时候刮板会偏离要刮掉的表面。在切削加工术语中，这种偏离的趋势称为刀具偏转。

一般来说，机床和刀具刚性越差，刀具偏转越有可能发生。虽然小的刀具偏转可能不足以引起大问题，但当零件的加工精度要求很高时则会影响很大。

在前面讨论的有关铣削工件轮廓右侧的示例中，即使使用直径精确为 1in 的立铣刀，在加工过程中仍然存在刀具偏转的可能。按照某表面的预期公差，大的偏离量可能足以导致零件超差。如果是这种情况，再使用固定的刀具轨迹中心线坐标，则意味着必须重新编程以减少铣削偏差。

还要注意的是，随着刀具磨损，刀具偏转发生的可能性将增加。这意味着锋利的刀具比钝的刀具有更少的偏转量。这时如果仍使用刀具轨迹中心线坐标编程，刀具寿命期内的偏转程度的变化则可能会影响加工精度。

3.4.3 复杂型面加工

对于简单零件，可以使用刀具轨迹中心线坐标编程。在前面给出的铣削工件轮廓右侧的简单示例中，将刀具半径简单地加或减到要加工的零件的方形表面相对容易。然而，随着铣削加工零件轮廓越来越复杂，立铣刀轨迹中心线坐标的计算也变得更加复杂。如果加工曲面和圆弧面，很多时候计算工件上的坐标会很困难，更不用说计算刀具轨迹中心线坐标了。

当数控程序员计算刀具轨迹中心线坐标（图 3.11）时，会发现计算量很大。

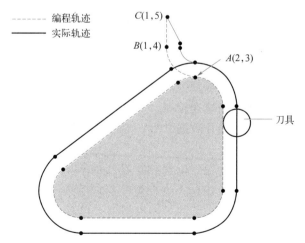

图 3.11 计算刀具轨迹中心线坐标示意图

使用刀具半径补偿的原因还与粗加工有关。当加工一个复杂的轮廓时，工件坐标很难得到，更不用说刀具轨迹中心线坐标了。除此之外，在粗加工过程中，整个轮廓表面有固定的精加工余量，没有刀具半径补偿，数控程序员完成的工作量将大大增加，其不仅要在精加工过程中计算立铣刀的轨迹中心线坐标，还要在粗加工过程中计算立铣刀的轨迹中心线坐标。

3.5 CNC系统的加减速控制

CNC机床进给速度的平稳性和冲击振荡控制与加工精度密切相关。在实际加工中，进给速度的突变产生冲击、振荡或超程、失步等动态误差，严重影响加工精度。研究表明，加减速过程中进给速度的不平稳是造成机床冲击、振荡的主要原因。

数控程序的数据读取过程，译码（解释）程序起核心作用，译码程序的主要功能是将用文本格式（ASCII）表达的加工程序，以程序段为单位转换为后续程序要求的数据格式，并存储在内存中供插补使用。通常，CNC系统发出与译码数据相关的指令，译码程序逐一读取指令，重复进行，直到译码结束。如果译码的时间大于加工指令执行时间，则机床将等待下一个程序段的译码，这样就造成停工现象。因此，为了保证加工过程的连续性，使用内部数据缓冲寄存区来临时存储译码数据，在当前加工程序段插补完成以前，必须完成下一个程序段的译码。

插补器顺序读取来自内部数据缓冲寄存区的数据，计算坐标轴联动过程中各轴单位时间位置和速度，并将结果存储在加减速控制器的FIFO（先进先出）缓冲器中。内插器的作用是计算每个轴的单位时间的位置和速度。直线内插器和圆弧内插器通常用于CNC系统。CNC系统中最典型的插补算法为直线插补和圆弧插补，有些CNC系统采用抛物线插补和样条插补。插补器根据路径类型（如直线、圆弧、抛物线和样条曲线）生成与路径数据相对应的脉冲，并将脉冲发送到FIFO缓冲器。脉冲的数量决定位置，脉冲的频率决定速度。脉冲当量决定了CNC伺服系统的位移精度。

例如，如果某个轴运动控制脉冲当量为0.002mm，则CNC伺服系统的位移精度是0.002mm。如果CNC系统以每秒8333个脉冲的脉冲频率发出25000个脉冲，相当于以1m/min的速度使工件移动50mm。

通过插补器生成数据执行位置控制，当工件运动开始和停止时，会发生较大的机械振动和冲击。为了防止机械振动和冲击发生，在向位置控制器发送插补数据之前，对加减速进行控制，这种方法称为插补前加减速控制。

同理，在向位置控制器发送插补数据之后，对加减速进行控制，这种方法称为插补后加减速控制。

来自加减速控制器的数据被发送到位置控制器，并且以固定时间间隔发送数据，执行位置控制。位置控制器通常为PID控制，它向电动机伺服系统发出速度指令，实现指令位置和编码器实际位置的跟随误差最小。如果用模拟信号控制，则噪声干扰较大。

CNC系统的进给速度控制

加速、减速和冲击控制由CNC系统或编程实现。

线性加减速规律如图3.12所示。

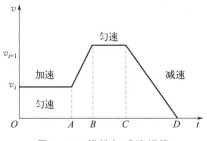

图 3.12　线性加减速规律

指数加减速规律如图 3.13 所示。

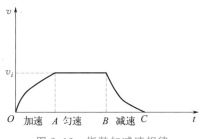

图 3.13　指数加减速规律

3.6　CNC 系统中的 PLC

数控内核（numerical control kernel，NCK）模块是 CNC 系统的核心，具有译码、插补计算、位置控制和基于译码程序的误差补偿等功能，通过机床伺服系统实现零件加工。可编程逻辑控制器（programmable logic control，PLC）在 CNC 系统中具有实现换刀、主轴转速控制、上下料和更换工件、I/O 开关量信号处理等除伺服控制外的大多数控制功能。

PLC 在机床和工业控制中用于顺序控制。过去，逻辑控制器是通过使用由继电器、计数器、定时器和电路组成的硬件来实现的，因此，它被认为是一种基于硬件的逻辑控制器。现在的 PLC 由半导体集成电路组成，其中包括微处理器和存储器。PLC 能够执行逻辑运算、计数、定时功能与算术操作等面向用户的指令。

PLC 的优点如下。

(1) 适应性强：使同一设备通过改变程序而改变生产过程成为可能。

(2) 扩展性好：通过添加模块或修改程序方便实现功能扩展。

(3) 经济实用：设计简单、成本低、可靠性高、维护方便。

(4) 体积小：与继电器控制箱相比体积更小，很容易装入机械内部。

(5) 可靠性高：采用半导体集成电路，故障率大幅降低。

(6) 性能优越：能够实现数学运算与数据处理等高级功能。

PLC 的结构与功能如图 3.14 所示。

CNC 系统的 PLC（图 3.15）与普通 PLC 类似，只是多了一个辅助控制器来协助完成

NCK 模块功能。

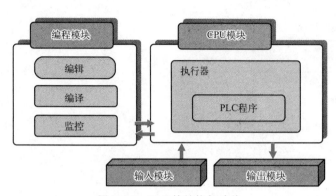

图 3.14 PLC 的结构与功能

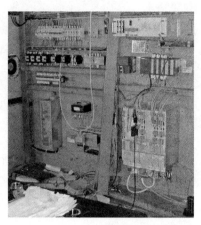

图 3.15 CNC 系统的 PLC

PLC 需要具备以下功能。

(1) 与 NCK 模块通信的专门电路。

(2) 支持高速通信的双端口 RAM。

(3) 与 NCK 模块高速通信时的数据存储器。

(4) 用于刀架等的高速控制的高速输入模块。

实际上，各 CNC 系统和 PLC 制造厂家都设计了自己的 PLC 编程语言，因此会造成维护性差和用户培训困难等问题。为了解决这些问题，国际电工委员会（IEC）建立了标准的 PLC 语言（IEC 61131-3），并进行推广。IEC 61131-3 定义了以下五种 PLC 标准编程语言。

(1) 结构化文本（ST）。

(2) 功能块图（FBD）。

(3) 顺序功能图（SFC）。

(4) 梯形图（LD）。

(5) 指令表（IL）。

数控程序员需要设计用于解释和执行 PLC 语言的应用程序，而用户只需按以上五种 PLC 标准编程语言编写 PLC 程序。

PLC 是一种用于机电过程自动化（如自动装配线、游乐设施或照明设备的控制）的数字计算机，广泛应用于 CNC 机床等工业控制领域。

等效继电器电路及其梯形图如图 3.16 所示。

梯形图的基本符号定义如下。

┤├ 常开（NO）触点：常开（NO）触点是一种电触点，当控制线圈（或继电器）断电时，它保持在断开位置（断开）。当线圈通电时，触点闭合，允许电流流过。

┤/├ 常闭（NC）触点：常闭（NC）触点是一种电触点，当控制线圈（或继电器）断电（断开）时，它保持闭合（连接）。当线圈通电时，触点断开，中断电流。

○ 输出（或线圈）：如果从左到右的连续触点都闭合，则输出得电。如果没有从左到

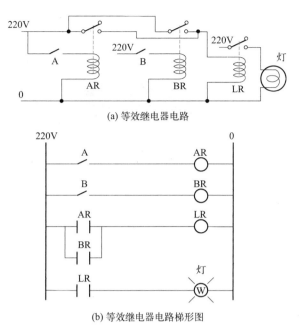

图 3.16　等效继电器电路及其梯形图

右的连续触点闭合，则输出断电。

在 PLC 中，CNC 装置被划分为多个功能模块，其硬件和软件采用模块化设计方法，即每个功能模块由相同尺寸的印制电路组成，各功能模块的控制软件也是模块化的。因此，用户可以通过将所选功能模块组合到主板的扩展槽中来建立自己的 CNC 装置。数控功能模块主要包括 CNC 控制模块、位置控制卡、PLC 模块、图形显示卡和通信卡等。

本 章 小 结

通过本章学习，要求学生了解 CNC 系统的组成，掌握 CNC 装置结构和工作过程，理解插补的基本原理和刀具半径补偿原理，通过常用插补算法拓展数控插补原理，通过刀具半径补偿原理掌握假换刀和刀具磨损补偿的理论基础，通过了解 PLC 拓展数控参数设置和硬件调试，掌握数控技术常用词汇，如 CNC unit、tool radius compensation、position controller、MCU、CLU、GPU、ROM、RAM、point-by-point comparison method interpolation、point-by-point comparison method circular arcs interpolation、tool length compensation、dimensional tool offsets、tool nose radius compensation、error compensation of tool、acceleration/deceleration controller、the linear law of acceleration and deceleration、the exponential curve law of acceleration and deceleration、NCK、PLC 等。

本章的知识和能力图谱如图 3.17 所示。

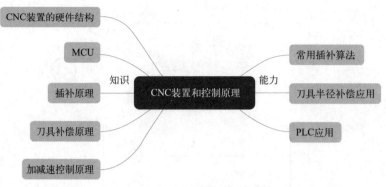

图 3.17 CNC 装置和控制原理知识和能力图谱

第 4 章

CNC 机床的伺服系统和位置检测装置

教学目的及要求

了解 CNC 机床伺服系统概述；

了解伺服电动机；

掌握位置检测装置。

4.1 CNC 机床伺服系统概述

4.1.1 伺服系统的概念

伺服系统是 CNC 系统的主要子系统。如果说 CNC 装置是 CNC 机床的"大脑"，是发布命令的"指挥机构"，那么伺服系统就是 CNC 机床的"四肢"，是一种"执行机构"。伺服系统是 CNC 装置和机床本体的联系环节，其性能直接影响 CNC 机床的精度、速度和可靠性。

典型的三轴 CNC 机床的伺服系统如图 4.1 所示。CNC 机床由机床本体、电力电子器件和 CNC 装置组成。机床本体主要由主轴传动装置、进给驱动装置组成。主轴电动机和进给驱动电动机及其伺服放大器、电源和限位开关是电力电子器件的一部分。CNC 装置由计算机系统、伺服驱动机构的位置传感器和速度传感器等组成。

本章介绍的伺服系统由机械传动机构、电力电子器件和部分数控单元组成。在伺服系统中，CNC 系统接收数字指令，处理并放大成强电驱动信号，以驱动电动机。当运动部件移动时，传感器检测执行机构的速度和位置，通过数控单元实时处理，由速度和位置反馈信号来实现进给控制和刀具轨迹的闭环控制。

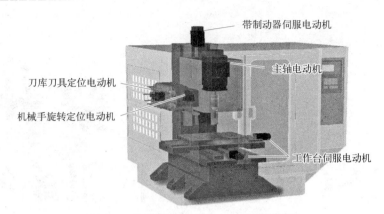

图 4.1 典型的三轴 CNC 机床的伺服系统

4.1.2 伺服系统的要求

1. 定位准确性好

CNC 系统执行精密加工或其他切削加工,其位置控制系统必须具有很好的定位准确性。定位准确性是在最坏的工况下定义的,这时期望的目标点位于相邻的两个坐标点中间。由于工作台移动目标是这两个坐标点之一,因此工作台的最终位置会有误差。这是最大定位误差,如果目标接近任一坐标点,则工作台会移动到较近的一个点,并且误差会较小,因此在这种最坏的情况下定义定位准确性是合适的。

2. 重复性好

重复性是指位置控制系统返回编程给定坐标点的能力。重复性可以用系统重复达到给定坐标点时的定位误差来衡量。

位置控制系统传动机构定位误差的分布符合正态分布曲线的统计规律。

3. 动态响应快、工作稳定

动态响应是指能按照指令脉冲频繁快速启动、停止、正反转,并且快速消除负载干扰。动态响应速度是伺服系统动态品质的重要指标,它反映了系统的跟踪精度。

稳定性是指系统的输出消除振荡的能力,即在给定输入或外部干扰作用下,在短时间的调节下,能够达到新的平衡状态或恢复到原来的平衡状态。系统的稳定性直接影响数控加工精度和零件表面粗糙度。

4. 调速范围宽

调速范围是指调速系统实际可以达到的最高转速与最低转速的比值,一般应大于 10000∶1,并要求保持稳定。

5. 低速大转矩

进给伺服控制属于恒转矩控制,应在整个调速范围内保持大转矩。

6. 没有累积误差

累积误差是在一系列测量或计算过程中逐渐累加的误差。确保没有累积误差对于保持

CNC 系统的精度和准确度至关重要。

4.2 伺服控制

4.2.1 伺服电动机的特点

机床（如车削机床和加工中心）在低速范围内的强力切削需要高转矩，在高速范围内的快速移动需要高速度（低速大转矩、高速大功率）。此外，对于循环加工、作业时间很短的机床（如冲床和高速攻螺纹机）需要具有小惯性和高可靠性的电动机。

CNC 机床伺服电动机的特点如下。

(1) 负载转矩大。
(2) 指令的动态响应速度大。
(3) 具有良好的加减速性能。
(4) 调速范围宽。
(5) 速度可控性好。
(6) 连续运行时间长。
(7) 能够频繁加减速。
(8) 具有高分辨率，以便在单步运动情况下产生足够的转矩。
(9) 旋转灵活、回转精度高。
(10) 制动转矩大。
(11) 可靠性高、使用寿命长。
(12) 维护简单。

伺服电动机设计时要满足以上特性。伺服电动机有直流伺服电动机、同步交流伺服电动机和感应式交流伺服电动机等，如图 4.2 所示。

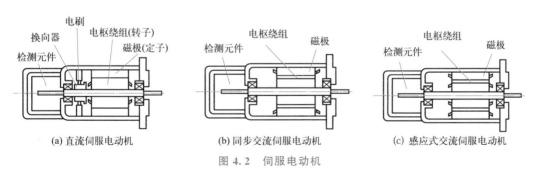

(a) 直流伺服电动机　　(b) 同步交流伺服电动机　　(c) 感应式交流伺服电动机

图 4.2　伺服电动机

伺服驱动系统是 CNC 机床的重要组成部分，其精度和重复性在很大程度上取决于驱动系统的特性和性能，要求驱动系统必须根据编程指令准确响应。大型 CNC 机床有时用液压马达驱动，但更多使用电动机驱动。电动机直接或通过齿轮箱连接到机床丝杠上，以带动机床导轨或主轴运动。常用的电动机有三类：步进电动机、直流伺服电动机和交流伺服电动机。

1. 步进电动机

步进电动机是一种将电脉冲数字信号转换为电动机轴离散旋转运动的机电设备。对 CNC 机床来说，步进电动机是最简单的驱动装置，因为它可以以将电脉冲数字信号转换为旋转运动。步进电动机的控制十分方便，不需要任何 A/D 转换器或反馈装置，非常适合开环控制系统。

但是，由于步进电动机具有速度小、转矩小、分辨率低、过载时容易失步等缺点，在 CNC 机床上应用受限。步进电动机的常见应用有计算机软驱和硬盘驱动器、菊轮打印机、数控电火花线切割机。

步进电动机有以下四种主要类型。

（1）永磁式步进电动机（可细分为听罐式永磁步进电动机和混合式永磁步进电动机，听罐式永磁步进电动机成本低廉，混合式永磁步进电动机具有高性能的轴承、较小的步距角和高功率密度）。

（2）混合式步进电动机。

（3）反应式步进电动机。

（4）拉维特型步进电动机。

简易步进电动机（单极）的步序如图 4.3 所示。

1～4—绕组。

图 4.3 简单步进电动机（单极）的步序

第一拍——绕组 1 通电，电磁力吸引相邻转子的齿对齐（使磁阻最小）。

第二拍——绕组 1 断电，绕组 2 通电，电磁力吸引相邻转子的齿对齐（使磁阻最小），旋转 3.6°。

第三拍——绕组 3 通电，再次旋转 3.6°。

第四拍——绕组 4 通电，再次旋转 3.6°。绕组 1 再次通电时，转子将旋转一个齿位。由于在本例中有 25 个齿，因此需要 100 步才能旋转一周。

步距角的计算公式为

$$\theta = \frac{360}{mzk} \tag{4.1}$$

式中，m——绕组相数；

z——转子齿数；

k——单拍 $k=1$，双拍 $k=2$。

CNC 装置：按控制要求发出指令脉冲，指令脉冲个数代表移动距离，脉冲频率代表移动速度，每发出一个脉冲，步进电动机旋转一个特定角度（步距角）。

环形分配器：根据指令方向，依次产生步进电动机的各相的通电步骤。环形分配器可以分为硬件环形分配器和软件环形分配器。

放大器：放大器用于放大环形分配的各相指令环，产生步进电动机的各相驱动电流。

步进电动机的工作原理如下。

步进电动机通过对步进电动机绕组电流的顺序控制来执行脉冲序列的转换。一般来说，每发出一个脉冲，步进电动机旋转一个特定角度（步距角），这种精确控制由步进电动机驱动器完成，步进电动机驱动器控制步进电动机的速度和位置。步进电动机按脉冲增加数将数字信号转换为精确的增量旋转步数，整个过程不需要反馈装置（如测速仪或编码器）。由于步进电动机及其驱动器是开环控制系统，因此没有伺服电动机系统中常见的反馈电路相移和由此产生的不稳定问题。

通过脉冲控制，步进电动机可以实现双向、同步控制，完成快速加速、启动/停止，并且易与其他数字控制系统集成。步进电动机具有转子转动惯量小、无漂移、定位误差不累积等特点，是许多运动控制经济有效的解决方案。步进电动机一般用于开环控制系统，没有反馈装置，有时可与直流伺服系统的性能相媲美。如上所述，与步进电动机相关的唯一误差是步距角精度（步进电动机每旋转一个步距角的实际值与理论值的误差用百分比表示）的非累积定位误差。通常，步进电动机的精度为3‰～5‰。

步进电动机和伺服电动机的主要区别在于电动机的结构和控制方式。步进电动机使用50～100齿的转子（没有电刷）和多磁极定子，步进电动机可以准确地在多个极点之间精确地移动。伺服电动机的转子与定子结构更复杂，而且大多数情况下配备了编码器或其他反馈装置。步进电动机把电脉冲信号转换成机械角位移，不需要使用反馈装置，而伺服电动机需要使用反馈装置来计算电动机控制电流。

步进电动机和伺服电动机的性能有所差异主要因为其设计原理不一样。步进电动机的极数明显多于伺服电动机的极数。与伺服电动机相比，步进电动机旋转一周需要更多绕组电流脉冲，因此，步进电动机在高速下的转矩性能差。与相同尺寸的伺服电动机相比，步进电动机的极数多，有利于在低速下提供更大的转矩，提高步进电动机的驱动电压可以减少步进电动机在较高转速下转矩降低的影响。伺服电动机和步进电动机的优缺点比较见表4-1。

表4-1 伺服电动机和步进电动机的优缺点比较

项目	伺服电动机	步进电动机
优点	高速转矩大	低速转矩大
	发热低	驱动控制器成本低
		较好的位置精度和运动重复性
		定位稳定
		静力矩（保持转矩）
		易于维护，可靠性高
		适用性广，易于开环控制或闭环控制
缺点	驱动控制器结构复杂	无反馈时发热严重
	低速转矩小	高速转矩小
	成本高	控制不当容易产生共振

步进电动机的使用特性如下。

(1) 步距角 (θ) 和步距误差。

步距角是转子在两个相邻脉冲之间转过的角度（每输入一个脉冲电信号转子转过的角度）。一般来说，步距角越小，控制精度越高。

步距误差（单步运行时转子转角的误差）直接影响执行部件的定位精度。步进电动机单相通电时，步距误差取决于定子和转子的分齿精度及各相定子的错位角度的精度；步进电动机多相通电时，不仅与上述因素有关，而且与各相电流大小、磁路性能有关。

(2) 最高起动频率（f_{st}）。

空载时，步进电动机由静止状态突然起动，并进入不失步正常运行，所允许的起动频率最高值为最高起动频率。当起动频率高于该值时，步进电动机便不能正常运行。最高起动频率与步进电动机的负载惯量有关。

(3) 最高工作频率（f_{max}）。

步进电动机起动后，其运行速度根据脉冲频率连续上升而不失步运行时，步进频率的最大值称为最高工作频率。最高工作频率也与负载有关。在相同负载下，最高工作频率远大于最高起动频率。

(4) 转矩频率特性。

在连续运行状态下，步进电动机的电磁转矩随脉冲频率的升高而急剧下降。电磁转矩与脉冲频率之间的这种关系称为转矩频率特性（简称矩频特性）。

选择步进电动机时要考虑以下关键参数。

(1) 机械运动参数。

(2) 步进电动机转速。

(3) 负载（转矩）。

(4) 步进驱动模式（整步驱动、半步驱动、细分驱动）。

(5) 绕组配置（相数）。

控制系统设计者在为特定应用选择配套的步进电动机、驱动器、控制器之前，需要了解机械运动参数、负载特性、耦合问题和电气要求。在确定步进电动机最优解决方案时，需要很好地考虑这些关键因素。设计者应调整控制系统各部件的特性，以满足设计要求，需要考虑的因素包括步进电动机、驱动器和电源的选择，以及机械传动机构（如齿轮传动或减轻负载）的替代方案。

图 4.4　步进电动机的铭牌示例

步进电动机的铭牌示例如图 4.4 所示。其主要包括额定电压、线圈电阻和步距角。步距角参数对于配置软件和正确控制驱动机构至关重要。对于一个三轴数控机床，至少 X 轴和 Y 轴要选用相同的电动机。

2. 直流伺服电动机

直流伺服电动机是 CNC 机床中常见的进给电动机类型。其工作原理基于电枢绕组在通电磁场中的旋转。电枢绕组连接到换向器，换向器安装在转子轴的绝缘铜触头上，直流电通过连接转轴端的电刷传递到换向器。电动机转速的变化是通过改变电枢电压来实现的，而转矩的控制是通过控制电枢电流来实现的。为了获得必要的动态特性，闭环控制系统中装有传感器，以获得速度和位置反馈信号。

直流伺服电动机［图 4.2(a)］的定子由一个圆柱形机壳组成，该机壳起到磁通和机械支撑的作用，磁极附着在机壳内部。转子由电枢轴和电刷组成。换向器和转子铁芯固定在电动机转轴上，电枢线圈在转子铁芯中。电刷通常由碳材料制成，通过换向器将直流电流传递到电枢线圈中。在转轴后端，转子中装有一个检测转速的传感器。一般来说，常用的传感器是光电编码器或测速发电机。

在直流伺服电动机中，由于转矩与电流成正比，因此用简单的电路就可以设计出控制器。限制功率输出的主要因素是电动机内部由于电流过大产生的发热。因此，有效的散热是产生高转矩的关键。直流伺服电动机调速范围广，价格低。但是，与电刷的摩擦会导致机械损耗和噪声，故要定期维护保养电刷。

3. 交流伺服电动机

在交流伺服电动机（图4.5）中，转子是永磁体，而定子是三相绕组。转子的转速等于旋转磁场的转速。

交流伺服电动机正在逐渐取代直流伺服电动机。其主要原因是交流伺服电动机中没有换向器或电刷、维护简单。此外，交流伺服电动机具有更小的比功率和更大的响应速度。

图4.5 交流伺服电动机实物图

交流伺服电动机的结构如图4.6所示。

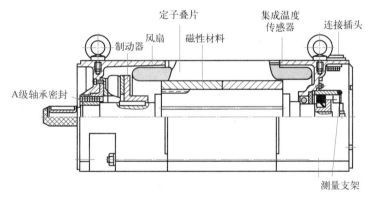

图4.6 交流伺服电动机的结构

（1）同步交流伺服电动机。

同步交流伺服电动机的定子由圆柱形机壳、电枢铁芯和电枢线圈组成。电枢铁芯位于圆柱形机壳中，电枢线圈绕在电枢铁芯周围。电枢线圈的末端与一根导线相连，并由导线提供电流。转子由转轴和永磁体组成，永磁体连接在转轴的外侧。

在同步交流伺服电动机中，磁铁附在转子上，电枢线圈绕在定子上，与直流伺服电动机不同，同步交流伺服电动机的定子不需要提供直流电，因其结构特点而被称为无刷伺服电动机。这种结构方便外部直接冷却定子铁芯，所以温升较低。另外，同步交流伺服电动机由于没有换向产生的火花，因此电动机的最高速度不受限制，可以获得高速范围内转矩大的良好特性。由于没有电刷，因此同步交流伺服电动机可以长时间运行而无须维护。

与直流伺服电动机一样，同步交流伺服电动机使用光电编码器或旋转变压器作为转速传感器。此外，铁氧体磁体或稀土磁体内置在转子中，起到磁场系统励磁绕组的作用。在同步交流伺服电动机中,由于电枢电流与转矩成线性比例,因此在紧急停止时制动快,能实现动态制动。

由于同步交流伺服电动机采用了永磁体,结构复杂,因此需要对转子位置进行检测。电枢电流为高频电流,高频电流是产生转矩脉动和振动的原因。

(2) 感应式交流伺服电动机。

感应式交流伺服电动机的结构与普通感应电动机相同。如果多相交流电通过定子绕组,则在转子绕组中产生感应电流,感应电流产生转矩。在感应式交流伺服电动机中,定子由机壳、定子铁芯、电枢绕组和引线等组成,转子由转轴和由导体构成的转子铁芯组成。

感应式交流伺服电动机结构简单,不需要检测转子与定子之间的相对位置。与普通交流伺服电动机不同的是,感应式交换伺服电动机在制动过程中存在励磁电流,因此会发生热损失,并且不能进行动态制动。

三种伺服电动机的优缺点比较见表 4-2。

表 4-2　三种伺服电动机的优缺点比较

项目	直流伺服电动机	同步交流伺服电动机	感应式交流伺服电动机
优点	成本低 调速范围宽 控制简单	无刷 易停	结构简单 不需要传感器
缺点	电刷发热、磨损 噪声大 需要位置检测装置	结构复杂 转矩脉动 振动 需要位置检测装置	无法动态制动 有热损失
尺寸	小型	中、小型	中、大型
传感器	不需要	需要光电编码器、旋转变压器	不需要
寿命	与电刷使用寿命有关	与轴承使用寿命有关	与轴承使用寿命有关
高速性能	差	一般	好
抗干扰性	差	好	好
永磁铁	有	有	无

4. 直线电动机

直线电动机(图 4.7)可以看成是一台旋转电动机按径向剖开,并展成平面。直线电动机产生推力的原理与旋转电动机产生转矩的原理相同。

通过线圈组件和永磁组件之间的电磁相互作用,电能被转换为直线运动的机械能。由于电动机的运动是线性的而不是旋转的,因此称之为直线电动机。直线电动机具有速度大、精度高、响应快等优点。20 世纪 80 年代,机床制造厂家开始推出直线电动机驱动的 CNC 机床。

图 4.7 直线电动机

在直线电动机的各种设计中,永磁无刷电动机具有力密度高、极限速度高和力常数稳定等优点。常用的无刷换向器具有维护方便、可靠性高、平稳性好等优点。

有铁芯的直线电动机[图 4.8(a)]相当于传统的无刷旋转电动机按径向剖开,并展成平面。展开的转子是一个固定板,在铁板背后贴有磁铁,展开的定子是一个移动线圈组件,由绕在叠层铁芯上的线圈组成。线圈通常以传统的三相排列方式连接,换相通过霍尔效应传感器或正弦信号实现。有铁芯的直线电动机具有效率高、推力大等特点。

无铁芯的直线电动机[图 4.8(b)]由一个固定的 U 形槽组成,槽内填充永磁体,永磁体沿两个内立面均匀分布。次级线圈在两排相对的磁铁之间移动。换相通过霍尔效应传感器或正弦信号实现。由于无铁芯的直线电动机在外壳部件之间没有产生引力,也不会产生锁定力,因此无铁芯的直线电动机具有铁芯质量小、电感低、无齿槽、运动平滑等优点。

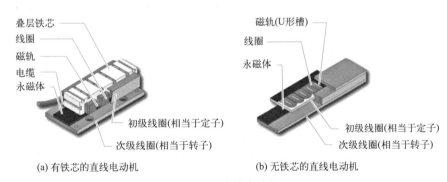

(a) 有铁芯的直线电动机　　(b) 无铁芯的直线电动机

图 4.8 直线电动机

4.2.2 伺服控制

1. 速度控制

速度控制通过比较测量电动机转速与指令信号,将信号反馈给速度控制系统,从而获得速度值来实现。通常,角速度由转子变化角速度对时间的求导得到。其他相关的各驱动器信号包括磁通量和转矩控制系统。

2. 速度反馈装置

电动机的实际速度可以用安装在电动机轴端的测速发电机（图 4.9）产生的电压来测量。直流测速发电机本质上是一个小型发电机，产生与转速成比例的输出电压，将产生的电压与相对应的速度的指令电压进行比较，其差值可以用来修正电动机速度，以消除误差。

测速发电机包括交流测速发电机和直流测速发电机，其输出电压与旋转电动机轴的转速成比例，因此可用来测量转速和旋转方向。

图 4.9 测速发电机

3. 位置控制

位置控制要求具有非常好的可操作性，电动机通过转动一个指定角度来完成起动、运行和停止。位置控制驱动器广泛应用于制造业、物料搬运设备、包装系统、加工流水线中。测量角位置的反馈传感器通常用编码器。伺服系统中常采用位置环、速度环和电流环（分别被称为外环、中环和内环）三环结构来实现高精度控制。位置控制器可以是线性控制（如 PI 或 PID）、模糊控制、神经模糊控制，也可以是变结构控制（滑模控制）。

4.3 位置检测装置

位置控制闭环系统是将插补计算的理论位置与实际反馈位置相比，用其差值控制进给电动机。而实际反馈位置的采集由一些位置检测装置（如角度或直线位移检测装置）来完成反馈位置和速度信号等。

为了使 CNC 机床能够精确地工作，需要实时获得各坐标轴的位置和速度。常用的反馈装置有位置检测装置和速度检测装置。这里仅介绍位置检测装置。位置检测装置有用于直接位置测量的线性位移传感器和用于角位移或线性位移间接测量的旋转编码器。其中，线性位移传感器是安装在机床工作台上直接测量导轨实际位移的装置。这样，丝杠和电动机等的间隙不会在信号反馈时造成任何误差。与安装在丝杠或电动机上的其他测量装置相比，该装置被认为具有更高精度和更高成本。

4.3.1 位置检测装置要求

为了确定机床每个轴的导轨位置，需要安装位置检测装置来监视和比较当前实际位置与每个轴的指令位置，用其差值控制电动机运动。

位置伺服系统对位置检测装置有以下要求。

（1）高可靠性和高抗干扰性。

（2）满足精度和速度要求。

（3）使用、维护方便，适合机床运行环境。

（4）成本低。

4.3.2 位置检测系统分类

位置检测装置系统有如下分类。

(1) 直接检测系统和间接检测系统。

直接检测系统是将位置检测装置直接连接到机床导轨或工作台上进行检测的系统。它独立于丝杠或传动元件，消除了与丝杠或传动元件相关的误差；但其成本较高，必须在整个检测长度内保持精度，反馈回路中有机床的机械共振。直接检测系统可以用线性测量装置（如光栅或感应同步器）来实现。线性位移检测传感器是大多数机床制造厂家提供的一个高成本选项，但随着对更高精度部件的需求，消费者对更高产品质量有所追求，线性位移检测变得越来越受欢迎。

间接检测系统是将位置检测装置安装在待测量机床部件的丝杠或驱动元件上的系统。间接检测系统具有成本低、安装简单、维护方便等优点。其精度要求可以通过齿隙、丝杠螺距误差和精密设备来补偿。间接检测系统通常配有旋转位置检测装置，如旋转变压器和编码器。

(2) 增量式检测系统和绝对式检测系统。

在增量式检测系统中，每一次位移都是检测值的增量决定的，即与前一位置相比距离的增加量。增量式检测系统只能确定相对位置，先前的测量误差会影响所有后续测量（会产生累积误差）。

在绝对式检测系统中，所有坐标都是从固定的基准位置或坐标原点检测的，与上一次的坐标无关，电源故障后重启（中途启动）容易，累计误差小。在高分辨率检测场合，使用绝对式检测系统成本较高。

(3) 数字检测系统和模拟检测系统。

在数字检测系统中，位置或位移是用离散值来检测的。数字检测系统可以是增量式检测系统（如使用光栅和增量式编码器）或绝对式检测系统（如使用绝对式编码器）。在数字检测系统中，数字信号更容易存储，传输更可靠，无差错再现。数字检测系统不需要单独的 A/D 转换器。

在模拟检测系统中，位移被转换为易于检测的物理模拟量。该物理模拟量可能与原物理量完全不同。模拟位置检测装置可以是电感式传感器（如旋转变压器和感应同步器），其输出与位移成线性比例关系。

模拟检测系统的电压幅值信号代表轴的位置，例如，用距离的物理量来表示电压，则指定导轨的位移将由感应电压模拟。模拟设备通常简单、鲁棒性好、成本低。然而，由于模拟检测过程是连续的，因此直接显示运动部件的位置和状态比较困难。模拟位移检测装置适用于中、小型 CNC 机床。

4.3.3 常见的位置检测装置

1. 旋转变压器

旋转变压器（图 4.10）是一种位置传感器或变换器，用于测量其所连接轴的实时角位置。自第二次世界大战以来，旋转变压器及类似的同步分解器在军事装备中得到了广泛的

图 4.10 旋转变压器

应用,如用于测量和控制坦克和军舰炮塔的角度。旋转变压器的结构与小型电动机类似,有一个转子(连接到所要测量角位置的轴上)和一个定子(静止部分)来输出感应电动势。

所有旋转变压器产生的信号都与转子角度的正(余)弦成比例关系。由于每个转角位置都对应一个正(余)弦值,因此旋转变压器能在其转子旋转一周(360°)内提供绝对位置信号。这种绝对(相对于增量)测量位置的能力是旋转变压器相对于增量式编码器的主要优势之一。

2. 光栅

光栅(图 4.11)由两块透明的玻璃片组成,在玻璃片上精细刻出(光刻或气相沉积)大量紧密且分布均匀的平行刻痕,刻痕不透光,两刻痕之间的光滑部分可以透光,组成等宽且等间距的平行狭缝。两块玻璃片一长一短,通常,长玻璃片称为标尺光栅,固定在机床运动部件上;短玻璃片称为指示光栅,固定在机床固定部件上。

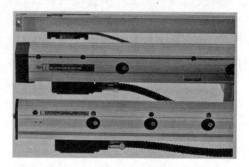

图 4.11 光栅

如图 4.12 所示,光栅测量系统由光源、准直透镜、玻璃光栅和光敏元件等组成。标尺光栅沿机床导轨行程安装,指示光栅覆盖在标尺光栅上(当光栅读数头相对于标尺光栅移动时,指示光栅便在标尺光栅上相对移动)。除标尺光栅外,其他装置均装在光栅读数头内。

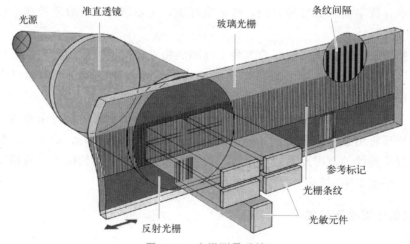

图 4.12 光栅测量系统

光栅的工作原理是莫尔条纹效应。指示光栅上的条纹和标尺光栅上的条纹相互交叉(光栅条纹之间的夹角为 θ)。在光源的照射下,交叉点附近的小区域内黑线重叠,形成黑色条纹，其他部分为明亮条纹。这种图案称为莫尔条纹（图4.13）。

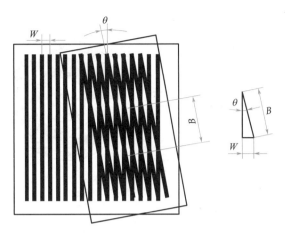

W—莫尔条纹宽度（节距）；B—栅距。

图 4.13 莫尔条纹

实际上，当导轨移动时，莫尔条纹会移动。莫尔条纹的移动方向与光栅的移动方向相互垂直，其移动方向取决于导轨向右（正）或向左（负）的移动方向。由图4.13可知，莫尔条纹宽度的计算公式为

$$W = \frac{B}{\sin\theta} \tag{4.2}$$

因为 θ 很小，当单位为弧度时，$\sin\theta \approx \theta$，所以

$$W \approx \frac{B}{\theta} \tag{4.3}$$

假设 $B=0.01$mm，$\theta=0.01$ 弧度，计算可得 $W=1$mm，即放大倍数为 100 倍。

3. 编码器

编码器（图4.14）用于测量角度位置或位移，它有以下两种类型。

（1）绝对式编码器：产生代码值，直接代表绝对位置。

（2）增量式编码器：产生数字信号，该数字信号以增量脉冲增减表示测量值。

绝对式旋转编码器（图4.15）是一种安装在电动机主轴或丝杠末端用于测量角位移的装置。由于绝对式旋转编码器不能直接测量直线位移，因此可能会因为丝杠、电动机等的间隙而产生误差。一般来说，该误差可以由机床制造厂家在机床校准过程中进行补偿。

绝对式编码器是一种角度测量装置。绝对式编码器在通电时可立即测量实际位置的值。

图 4.14 编码器

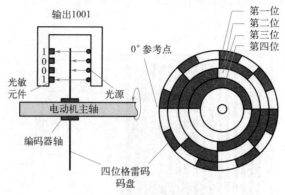

图 4.15 绝对式旋转编码器

绝对式编码器不需要计数装置,因为测量值直接来自物理编码刻度。在大多数情况下,绝对式编码器的输出是二进制编码或格雷码。

绝对式编码器有三种类型:接触绝对式编码器、光电绝对式编码器和电磁绝对式编码器。

增量式编码器(图 4.16)是一种旋转式脉冲发生器。增量式编码器是非常简单的位置检测装置,不能像绝对式编码器那样给出绝对位置。增量式编码器只能给出相对位置,通过计算脉冲数得到相对位移。增量式编码器包含一个带中心轴的透明码盘,码盘圆周上刻有相等间距(透明、不透明相间)的条纹。在码盘的一侧提供固定光源,另一侧放置光敏元件。

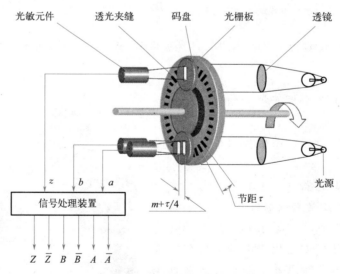

图 4.16 增量式编码器

当码盘旋转时,光线周期性地打在光敏元件上,当码盘缓慢沿边缘上升或下降会产生毫伏级的正弦输出信号,这一信号经放大和整形后得到方波信号。方波信号通过微分元件变为短周期脉冲。每一次光敏元件被遮挡都会产生一个脉冲信号。每转输出的脉冲数是码盘上线数的函数,称为分辨率。典型的码盘为 200~18000 线。

增量式编码器与驱动直线位移的动力装置的主轴联轴,通过计量输出脉冲数,由丝杠螺距齿轮比设定传动参数,脉冲计数可以用来确定工作台的直线位移。

本 章 小 结

通过本章学习,要求学生了解 CNC 机床伺服系统的组成及分类,理解常用位置检测装置的工作原理及应用特点,掌握步进电动机控制原理,通过对 CNC 机床伺服系统和常用位置检测装置的工作原理的学习拓展学生对 CNC 机床故障诊断、维修和维护的能力,掌握数控技术常用词汇,如 servo system、position measuring devices、repeatability、rapid response、stability、torque、cumulative error、driving motor、DC servo motor、synchronous-type AC servo motor、stepper motor、step angle、phase winding、start frequency、start torque-frequency、induction-type AC servo motor、linear motor、ironcore brushless linear motor、tacho-generator、analogue measurement system、resolver、optical grating、moire fringe effect、absolute encoder、incremental encoder 等。

本章的知识和能力图谱如图 4.17 所示。

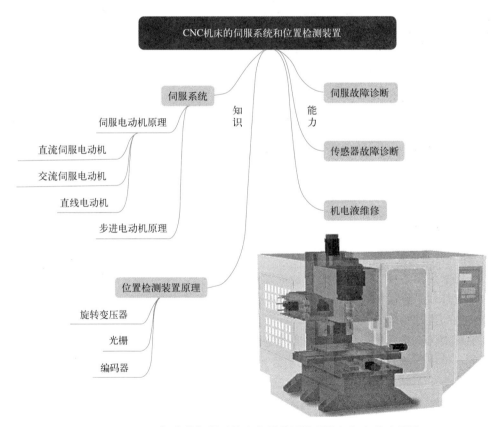

图 4.17 CNC 机床的伺服系统和位置检测装置的知识和能力图谱

第 5 章

CNC 机床的机械系统和刀具系统

教学目的及要求

了解 CNC 车床及 CNC 加工中心；
了解 CNC 机床的机械系统；
掌握 CNC 机床的刀具系统。

5.1 CNC 车床及 CNC 加工中心

机床的作用是切削多余的材料。用金属切削刀具把工件毛坯上的余量（预留的金属材料）切除，获得零件图所要求的零件形状和尺寸，保证其加工精度和表面粗糙度。为了满足这些要求，机床应具有以下性能。

(1) 能够牢固地夹紧工件和刀具。

(2) 功率足够，使刀具能够以合适的速度切削工件。

(3) 刀具和工件能够相对移动，并且必须以一定的精度进行控制，以确保工件表面粗糙度和理想的尺寸精度。

采用早期传统机床加工时，操作者必须站在机床前操作。采用 CNC 机床加工时，因为操作者不再控制机床的运动，所以不必站在机床前操作。在传统的机床上，只有大约 20％的时间用于去除材料加工，随着电子控制技术的应用，CNC 机床去除材料加工的实际时间增加到 80％，甚至更高，同时减少了移动刀具等的非加工时间。

5.1.1 CNC 车床

车床作为生产效率最高的机床之一，一直是加工回转体零件的主要手段。
CNC 车床具有以下特点。

(1) 刀具或工件可以移动。

(2) 刀具可以在1～5轴上加工。

(3) 大型CNC机床由MCU管理各种操作。

(4) 运动由电动机控制。

(5) 反馈由传感器完成。

(6) 刀库用于自动换刀。

CNC刀具具有以下特点。

(1) 大部分CNC刀具由高速钢、碳化钨或陶瓷制成。

(2) 设计刀具的目的是将多余材料直接从工件上移除。

(3) 部分刀具需要冷却（如油冷），以保护刀具并完成加工。

刀具轨迹、切削和输出运动具有以下特点。

(1) 刀具轨迹描述了刀具运动的路径。

(2) 运动可以描述为点对点控制、直线切削控制或轮廓控制。

(3) 切削速度是刀具工作的速率，如转速（r/min）。

(4) 进给速度是刀具相对工件移动的速率。

(5) 进给量和进给速度取决于所需的切削深度、工件材料和精加工精度。例如，加工较硬的材料需要较慢的进给速度和切削速度。

(6) 粗加工要比精加工去除更多的材料。

(7) 点定位（快速运动）允许刀具或工件在不进行加工时快速移动。

1. CNC车床的应用

CNC车床和普通车床都主要用于加工轴类、盘类等回转体零件。与普通车床相比，CNC车床的加工精度更高、加工质量更稳定、效率更高、适用性更强、工作强度更低，更适合加工复杂形状的轴类、盘类等回转体零件。

2. CNC车床的常规刀具和夹紧装置

在CNC车削加工中，成形刀具有小半径圆体车刀、非矩形切槽刀和螺纹车刀等。但是，这些成形车刀在实际中很少使用，只是偶尔使用或根本不使用。如有使用必要时，则应在工艺文件和加工程序清单中给出刀具的详细说明。

CNC车床与普通车床的夹紧式相似，主要有焊接式和机械夹固式。在CNC车床夹具中，常用的是自定心卡盘和四爪卡盘，在批量生产中也经常使用自动控制的液压夹具、电气夹具、气动夹具。另外，CNC车床对不同类别零件加工通常要采用与其相匹配的夹具，如轴类零件夹具和盘类零件夹具。

(1) 轴类零件夹具。

轴类零件夹具有自动夹紧卡盘、顶尖、自定心卡盘和可快速调心通用卡盘等。CNC车床加工轴类零件时，通过主轴顶尖与尾架顶尖之间的鸡心夹和安装在主轴上的拨盘带动工件旋转。这些夹具在外圆车削时可以传递足够大的转矩，以适应主轴的快速回转车削。

(2) 盘类零件夹具。

盘类零件夹具有可调爪形卡盘和快速可调卡盘。这些夹具适用于无尾架的CNC车床。

3. CNC 车床的刀具补偿

CNC 车床通常需要不同种类的刀具来加工零件。在加工同一零件时，每把刀具的刀尖位置是不一样的。刀具补偿就是测量各刀具的位置差，并将各刀具的刀尖统一在同一工件坐标系的某个夹紧位置上，这样每把刀具的刀尖轨迹都在同一工件坐标系。CNC 车床常用的刀具补偿方法如下。

（1）自动刀具补偿。

（2）试切对刀。试切对刀可用于 CNC 车床的闭环控制系统或开环控制系统。

（3）机床内置刀具补偿。机床内置刀具补偿用刀具接触固定的触头测量刀具偏置量并进行校正。

（4）通过对刀点实现刀具补偿。

根据 CNC 车床参数设置或调整机床各坐标轴的机械限位位置，在刀具补偿基准点上设置对应于刀具起点的参考点。这样，当 CNC 车床回到操作者设置的工件原点时，可以使刀尖回到起始位置。

4. CNC 车床的组成

CNC 车床是一种将工件毛坯夹在主轴上，并回转加工的机床，主要用于加工各种金属件的回转面和端面，以使用锋利的单刃刀具为主。CNC 车床利用转塔回转刀架紧固刀具并实现刀具的精确移动。

此外，转塔回转刀架具有全自动换刀功能。它用于快速取出旧刀具并将新刀具装入其加工位置。前置转塔回转刀架（刀架位于主轴和操作者之间）的主轴正转时刀尖朝上，后置转塔回转刀架（主轴位于刀架和操作者之间）的主轴正转时刀尖朝下。因此，同时配备前置转塔回转刀架和后置转塔回转刀架的 CNC 车床可以同时从工件上方和下方执行切削操作。

CNC 车床的组成（图 5.1）如下。

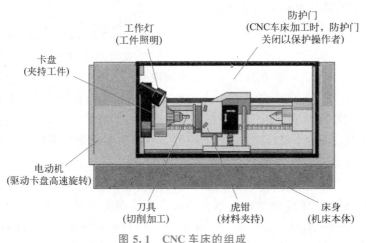

图 5.1　CNC 车床的组成

（1）虎钳。

虎钳装夹要加工零件的毛坯。毛坯必须牢固夹紧，当 CNC 车床加工时，毛坯不会从虎钳中高速抛出。简单地讲，虎钳就像一个将毛坯固定在正确位置的夹子。

（2）防护门。

防护门保护操作者的安全。当 CNC 车床加工工件时，切割下来的小碎片高速从母体射出，如果碎片击中操作者，那么将是很危险的。防护门将数控加工的危险区域完全隔离开来。

（3）卡盘。

卡盘连接在主轴上，其主要功能是夹紧工件。它装夹要加工零件的毛坯，毛坯定位要可靠，这样加工时毛坯不会被高速抛出。

（4）电动机。

电动机安装在机床内，驱动卡盘高速旋转。

（5）床身。

床身是机床的底座。通常，CNC 车床是用地脚螺栓固定的，这样它在加工时就不会因为机床的振动而移动。

床身是车床的基础支撑框架部件，它提供排屑通道。斜床身 CNC 车床的排屑更容易，它还配有便于刀架纵向滑动的导轨。CNC 车床的操作高度应符合人机工程学。

（6）刀具。

刀具通常由高性能的钢材制成，它的主要功能是使毛坯加工成形。

5.1.2 CNC 加工中心

铣削是常用的加工方法。金属去除是通过旋转、多切削刃铣刀的相对运动和工件的多轴联动实现的。铣削是一种断续切削，其中铣刀的周期性切入和切出运动实现了实际的金属去除和不连续的切屑生成。铣床一直是工业上用途广泛的机床，可以完成铣削、轮廓加工、齿轮加工、钻孔、镗孔和铰孔等，这些加工方法只是铣床能够完成的工作的一小部分。加工中心是铣床的一种。

所有的铣床，从紧凑的台式铣床到标准的立式升降台铣床和大型加工中心，其加工原理和切削用量相近。重要的切削用量如下。

（1）切削速度：刀具切削刃选定点相对于工件的速度。

（2）进给速度：刀具回转一周相对于工件的位移。

（3）轴向切削深度：刀具切削刃在工件未加工表面下方测量的距离。

（4）径向切削深度：工作面上被刀具切入的距离。

铣床的性能主要由电动机的功率确定，它决定主轴最高转速和主轴孔锥度。

加工中心是在各种非圆形表面或棱柱形表面上进行铣削和整体加工的机床。加工中心的独特之处是 ATC。ATC 能快速地将刀具从刀库移到主轴，然后放回去。大多数加工中心只能存储和处理 20～40 把刀具，个别加工中心刀库容量超过 200 把刀具。

铣削加工的核心是铣刀。铣刀是用于铣削加工的、具有一个或多个刀刃（切削刃）的旋转刀具，每个刀刃在每次切入和切出工件时只去除少量的材料。铣刀的种类繁多，最常用的铣刀是铣削平面的面铣刀，用于高速加工，直径从约 76.2mm 到约 609.6mm 不等，有的面铣刀能同时铣出一个与表面垂直的轴肩。

坡口加工、轴肩加工和铣槽无法用面铣刀完成。立铣刀（端铣刀）是一种圆柱表面和端面上都有切削刃的刀具，可以用于短槽、浅槽加工和一些轮廓周边的精加工。圆形槽铣刀或直槽铣刀更适合加工更长和更深的槽，这是因为立铣刀切削量较大时容易打偏。倒角和轮廓铣削常用特殊定制的立铣刀。

图 5.2　CNC 加工中心

在各种铣削加工中，一个关键的部件是工件夹具，它能够快速将新的工件或加工面定位到加工位置。长机床床身、托盘交换装置和多层编码夹具的结合使用为加工中心提供了极大的灵活性和高效性，可以在前一工件铣削时预先设置好工件的下一个加工位置。

CNC 加工中心（图 5.2）常包含两个非常有用的附件：一个是接触测量探头，从 CNC 加工中心切除工件之前，利用计算机软件对工件尺寸进行测量，探头与其他工具一起存放，以便快速调用；另一个是刀具预调仪，它允许技术人员在将刀具装入 CNC 加工中心刀库之前根据数控程序的要求设置刀具参数。

CNC 加工中心是配有 ATC 和刀库的 CNC 机床。CNC 加工中心是在 CNC 铣床的基础上增加刀库和回转工作台发展而来的，因此，CNC 加工中心具有铣削、镗孔、钻孔等功能。

CNC 加工中心的主要特点如下。

（1）工序集中，工件一次装夹可以完成多面加工。

（2）可配备自动分度装置（或回转工作台）和刀库系统。

（3）主轴转速自动控制。刀具进给量和运动轨迹与工件轮廓有关。

（4）其生产效率是普通 CNC 机床的 5～6 倍，特别适用于形状复杂、精度要求高、变化频繁的零件加工。

（5）操作者劳动强度低，但零件结构复杂，对操作人员的技术水平要求高。

（6）成本高。

虽然不同类型的 CNC 加工中心组成会有所不同，但是其主要组成部分都包括基础部件、主轴部件、CNC 系统、ATC 和附件。

其中，主轴部件可以垂直设置（立式），也可以水平设置（卧式），或者立卧复合设置，主轴垂直和水平设置可以转换。单面加工时通常首选垂直设置。使用回转工作台可以在不需要操作者干预的情况下完成加工单件多面或多件加工。带回转工作台的立式 CNC 加工中心有四个运动轴：三个轴是工作台的 X 轴、Y 轴、Z 轴（直线运动），第四个轴是回转台（工作台的回转运动）。

CNC 加工中心的分类如下。

（1）立式 CNC 加工中心：主轴轴线与工作台垂直设置的 CNC 运动加工中心，主要适用于板类零件的加工。回转工作台可安装在水平工作台上用于加工螺旋线。

（2）卧式 CNC 加工中心：主轴轴线与工作台平行设置的加工中心，装有带分度功能的回转工作台，斜置 3～5 轴坐标联动，主要适用于箱体类零件的加工。

（3）门式 CNC 加工中心：主轴通常与工作台垂直，并配有可更换的主轴头附件，主要适用于大尺寸或复杂形状的零件加工。

（4）立卧复合式 CNC 加工中心：利用立卧转换机构实现主轴的水平和垂直设置。工件一次装夹后，完成除定位外表面外的其他所有表面加工。它可以减小结构误差，避免二次装夹误差，生产效率高，成本低。

（5）带 ATC 和刀库的 CNC 加工中心：自动换刀通过刀库和换刀机构来完成，其优点是加工范围广。

（6）带转塔刀架的 CNC 加工中心：应用于小型立式 CNC 加工中心。

为了获得较高精度和可重复性，CNC 加工中心滑动导轨和驱动丝杠的设计与制造至关重要。滑动导轨通常需要高精加工，并涂以聚四氟乙烯（PTFE）和热塑性自润滑轴承材料等耐磨材料，以减少黏滑现象；采用大直径滚珠丝杠以消除齿隙和位移损失。

CNC 加工中心的其他设计要求有结构刚性大、床身悬伸短、换刀速度快等，这都有助于提高 CNC 加工中心的精度和可重复性。

5.2 CNC 机床的机械系统

CNC 机床的机械系统由驱动机构、导向机构、床身、立柱、工作台和切屑输送机构等组成。

1. 驱动机构

CNC 机床的驱动原理是将电动机提供的转矩转化为铣头的直线运动。带螺母的丝杠提供了一种简单而紧凑的能量传递方式。由于滚珠丝杠和丝杠螺母副摩擦力小、效率高，因此采用该传动结构。丝杠由于重载能力差和自锁复杂，不适用于 CNC 机床。因为滚珠丝杠运行效率高，直线移动速度驱动所需转矩小，所以优先选择滚珠丝杠作为 X 轴、Y 轴、Z 轴的驱动机构。另外，采用滚珠丝杠和丝杠螺母副作为动力传递，热稳定性好且可靠性高。

2. 导向机构

导轨支承龙门架和铣头的质量，同时在龙门架移动期间提供准确定位。直线导轨的圆柱表面硬化处理，并带滚柱衬套。但是，如果龙门架上的质量和负载产生的扭曲超过机床规定公差，则需要考虑更精密的轴或导轨结构。

3. 床身

床身（这里指龙门床身）包括立柱和工作台。床身要有足够的强度，才能支承龙门架的质量，并承受铣头及铣削过程中产生的切削力。

床身还要有足够的刚度，防止铣头加速度产生的静态力和动态力引起的扭曲。床身自重很关键，因为床身的质量对静态力和加速度都有影响。床身材料应满足三个要求——强度、刚度和质量，并具有良好的加工性能，成本也要低。通过对材料的综合考虑（强度、刚度、质量、加工性能和成本），确定用 HDPE（高密度聚乙烯）作为床身材料。在塑料材料中，HDPE 便宜，同时易加工，其强度和刚度也很好。

4. 立柱

在导轨上移动时，立柱（这里指龙门立柱）支承导轨和铣头的质量。立柱用 HDPE

材料制造，因为其必须轻量化，以减少加速过程中的惯性。

5. 工作台

工作台支承待加工的工件，并作为机床的底座。制造工作台的材料差别较大，加工和装配复杂。由于 HDPE 成本低且易加工，因此常用作工作台基材。在应用中，HDPE 的强度和刚度将经过合格检测。当机床沿工作台长度方向移动时，导轨支承的重量会造成工作台的侧面扭曲，这可能导致铣头的位移超过设计公差，HDPE 结构能够承受这些重量并保持加工公差。要解决扭曲问题，可以考虑采用复合结构系统，如采用铝加固 HDPE 底座侧面。

6. 切屑输送机构

切屑输送机构（图 5.3）是一种可移动的钢制平板，它清除机床上产生的金属切屑。在某些车床上，它与冷却液箱合为一体。切屑输送机构通常采用重型链板式结构，呈蛇形布置。

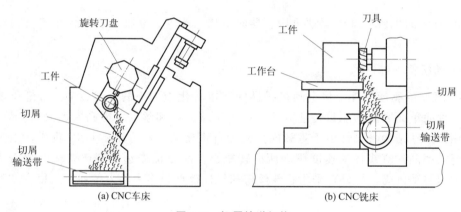

图 5.3 切屑输送机构

5.2.1 主轴设计

购买加工中心时，必须考虑主轴最高转速、主轴最大功率和主轴最大转矩等主轴参数。

主轴是加工中心的耐用件，加工中心制造厂家要确保主轴设计合理，并选用高质量部件保证主轴的使用寿命。CNC 车削中心的主轴设计如图 5.4 所示。

主轴需要的功率取决于加工的工件。购买立式加工中心时，其转矩、速度和功率是需要评估的重要指标，整体性能是否满意还要考虑其他因素，如主轴和机床的总体满意度、购买立式加工中心的 ROI 等。

1. 主轴的重要性

主轴在立式加工中心上似乎算不上什么核心部件，因为刀具用于切削工件，工作台是移动平台，运动控制系统决定精度和实现运动，其余工作就是编程，一般情况都会认为主轴不过是从伺服系统接收指令，通过电动机将刀具连接到主轴上，实现转动而已。

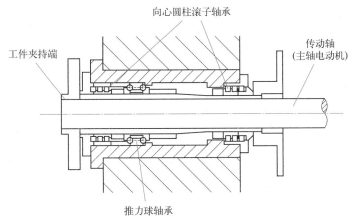

图 5.4 CNC 车削中心的主轴设计

上述看法也不能说是错的，主轴可能不是太复杂，也没那么"聪明美丽"，但它确实在努力工作，忍受了很多误解。主轴轴承出现磨损意味着主轴的结构和主轴内零件的质量影响了主轴的性能和使用寿命，主轴其实也是加工中心的核心部件。

零部件的质量不仅决定了主轴的使用寿命，还决定了主轴如何应对速度、转矩和振动。开始研究主轴技术时，主轴轴承部件常常是主轴质量讨论的中心议题。在加工中心主轴轴承部件研究中，热点问题包括主轴轴承的材料、类型、布置和润滑方式等。

2. 主轴轴承部件

在主轴轴承部件（图 5.5）中，滚珠在内外滚道之间滚动。滚珠轴承所用的材料影响主轴轴承温度、振动和使用寿命。与典型的滚珠轴承相比，混合式陶瓷轴承具有如下优势。

（1）质量小。

混合式陶瓷轴承的质量比滚珠轴承小 60%。当滚珠轴承工作时，特别是在高转速下，离心力将滚珠推到外圈，甚至造成滚珠变形，当滚珠轴承开始变形时，磨损加速，性能退化。而质量较轻的混合式陶瓷轴承在同等速度下工作没有受到太大的影响。事实上，据一家高速铣削主轴制造商提供的资料，

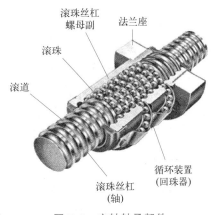

图 5.5 主轴轴承部件

与滚珠轴承相比，使用混合式陶瓷轴承可以在不牺牲轴承使用寿命的情况下将其转速提高 30%。

（2）消除冷焊合。

混合式陶瓷轴承没有了金属与金属的接触，陶瓷球不会与内外滚道发生反应（绝缘），消除了微焊合或冷焊合现象，以及相关的黏结剂和磨料磨损。当滚珠与滚道表面滑动导致表面磨损时，微小冷焊累积发生冷焊合。当滚珠轴承旋转时，冷焊合被撕裂，从而产生粗糙表面，导致发热和轴承故障。

(3) 低温运行。

由于陶瓷球有近乎完美的圆度,混合式陶瓷轴承的工作温度比滚珠轴承低得多,从而延长了混合式陶瓷轴承润滑油的使用寿命。

(4) 振动水平较低。

混合式陶瓷轴承的主轴具有较高的刚度和较高的固有频率,降低了振动敏感性,从而延长了混合式陶瓷轴承润滑油的使用寿命。

3. 轴承类型

角接触球轴承是设计高速主轴时常用的轴承,如图 5.6 所示。该类轴承能满足切削加工所需的精度、承载能力和速度要求。高精度滚珠装在内滚道和外滚道之间,可同时承受径向负荷和轴向负荷。

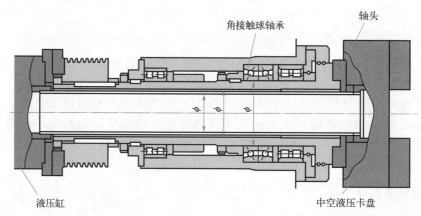

图 5.6 数控车床主轴的角接触球轴承

其他类型的轴承也常用于主轴设计中,如圆柱滚子轴承或圆锥滚子轴承。其中,或圆柱滚子轴承(滚柱轴承)比滚珠轴承具有更高的承载能力和更大的刚度,用于具有特定转速要求和环境要求的主轴设计中。通常,根据轴承负荷类型,主轴制造厂家在主轴的不同部位选用球轴承、圆锥滚子轴承或圆柱滚子轴承。

4. 轴承的润滑

轴承的正确润滑至关重要。机床制造厂家使用各种方法来保持轴承的适当润滑,如油雾、油气、喷油和脉冲油气润滑等。

如果主轴转速超过 18000r/min,就有必要使用轴承润滑系统,但它们增加了维护成本和主轴更换成本。此外,必须对这些润滑系统进行监控,以确保油、空气、油雾的配比正确。

永久润滑轴承是降低维护成本和更换成本的最佳选择。使用永久润滑轴承,用户不必为润滑而烦恼。它可以在主轴装配过程中填充,也可以由轴承厂家预先填充(永久润滑)。

5. 主轴类型

按传动方式不同,主轴分为带传动主轴(图 5.7)、齿轮传动主轴、直联传动主轴和电主轴。带传动主轴要确保定期维护和调试输送带,并以最大限度地降低维护成本。另外，输送带选用类型会影响机床的噪声控制，采用人字形输送带比采用其他输送带更安静，因为它可以分散气流，从而降低噪声。

若主轴采用齿轮传动，则机床成本较高，与带传动主轴相比，齿轮传动主轴噪声大、维护复杂。过去有一段时期，齿轮传动主轴比带传动主轴更受欢迎，但随着材料和输送带设计技术的发展，维护方便的带传动主轴越来越多地替代了齿轮传动主轴。

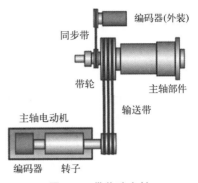

图 5.7 带传动主轴

直联传动主轴（也称直接传动主轴）的设计使主轴直接联轴到电动机上。直联传动主轴提高了工件表面加工精度，使操作更平稳、更安静。

电主轴的主轴与电动机转子合为一体。当需要更高的主轴转速（超过 16000r/min）时，通常使用电主轴。与带传动主轴相比，电主轴成本较高。

无论采用哪种类型的主轴，驱动主轴的电动机都尤为重要。带有两组绕组的电动机称为双绕组电动机，它具有更大的切削转矩和材料去除率；而单绕组电动机适用于低转矩和高转速的场合。

6. 主轴的"敌人"

主轴的两个主要"敌人"：杂物（如切屑和冷却液进入轴承系统）、发热。针对以上影响因素，通过主轴设计改进（或其他选项）以防护主轴。从历史数据看，主轴故障的常见原因是冷却液进入、冷凝、污染或切屑误入等。如果操作者不希望主轴发热，则应确保杂物不进入主轴。

在大多数情况下，杂物进入主轴是因为主轴密封失效。有必要了解机床制造厂家为保持主轴密封所采取的设计措施。空气净化系统使用迷宫式密封，并用正气压净化密封，这样可有效防止杂物进入。双循环空气净化系统有两个进出风口（通常上下布置），该净化设备也可以很好地防止杂物进入。

发热会导致钢材热膨胀，制造商应说明他们采取了什么措施来防止主轴头热伸长的影响，因为热伸长影响大部分机床 Y 轴和 Z 轴的位移。

换热器或冷却器常用于冷却主轴，并控制主轴热胀和头部热伸长，这类装置延长了主轴的使用寿命，减轻了主轴头热伸长的影响，通常应用在长加工循环或高效循环加工中。冷却器的选择取决于主轴的应用场合。对于高速主轴，可能需要研究热稳定系统，该系统使用带油冷却器的自动调节系统，可以根据需要自动冷却主轴。

刀具也影响主轴性能。使用不对称刀具、磨损刀具、太长的刀具都会影响主轴的使用寿命。

7. 冷却时的注意事项

发热会对刀具产生负面影响，因此需要冷却刀具。首先找出主轴是否带有冷却循环系统或可调冷却液喷嘴，如果有冷却循环系统，则要进一步明确喷嘴个数，以及它们是否可调。喷嘴越多越好，并且喷嘴的方向可调，以便实现刀具长度全方位冷却覆盖。

当主轴以 12000r/min 或更高的转速加工时，通常建议使用主轴中心出水冷却，高速加工时常使用定制刀具，以使刀具得到很好的保护。对于循环加工等应用场合，建议在较低转速下使用主轴中心出水冷却。此功能的成本根据主轴中心出水的压力和主轴设计的复杂程度而定。

8. 重置成本

更换立式加工中心的主轴要考虑成本。在决定购买全新的加工中心时，要确保能预见主轴需要更换的情况，需要知道更换成本，更换的主轴类型是否可用，以及停机时间是否影响生产。

5.2.2 导轨设计

设计床身时首先要考虑传统的相对运动部件——导轨系统，导轨系统由直线滑动轴承（运动件）和导向轴组件（承导件）组成，允许该轴承和轴组件沿导轨长度方向来回移动。考虑设计规范和尺寸要求，最合理的导轨设计思路应是：导轨能够以某种方式支承载荷，而且不会产生很大扭曲。例如，具有简单的钢制导向轴的导轨系统质量较小，近年来，为了提高导轨系统的性能，导轨的设计结构变化很大。

钢制导向轴导轨采用一种简单有效的直线运动设计。导向轴为轴向加载提供支承，以及承载由线性运动产生的力。

导轨系统使用导向轴及其承载系统来支承轴向荷载，承载线性运动的力。导向轴及其承载系统可以采用陶瓷材料，从而提高导轨的性能，减少振动，同时减少导向轴在加载情况下的挠度，提高导向轴的使用寿命。

导轨的设计千差万别，并在实际工程应用中不断完善，滚动导轨系统和滑动导轨系统分别如图 5.8 和图 5.9 所示。每个导轨系统，其零部件来自多个不同的供应商，包括陶瓷导轨、表面硬化处理钢制导向轴导轨系统。大多数导轨系统的钢件采用表面硬化处理，以与特定轴承相匹配。

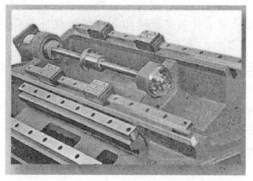

图 5.8 滚动导轨系统

图 5.9 滑动导轨系统

V 形导轨系统通过带 V 形槽的滚轮轴承在 V 形导轨面上滚动实现支承载荷和直线运动。V 形导轨系统可以设计得很复杂，通过 V 形导轨的顶部和底部悬挂在地面上，实现大跨距和长行程导向。

5.2.3 回转工作台设计

回转工作台是一种用于金属加工的精密工件定位装置。它使操作者能够在固定的（通常卧式或立式）轴上精确等分钻孔或加工工件。有的回转工作台允许使用分度盘进行分度操作，有的回转工作台装有分度板，完成分度盘完成不了的频繁定位夹紧。

图 5.10 所示为手动操作的回转工作台。在 CNC 机床控制下，工作台自带动力，相当于 CNC 铣床的第四轴。回转工作台有一个坚固的底座，底座上安装工作台或夹具装置。实际上，回转工作台是一个精密加工的圆盘，工件被夹紧在圆盘上（通常为此设置 T 形槽）。该圆盘可以自由旋转，用于分度，或在蜗杆（手轮）的控制下驱动与蜗轮相连的工作台的回转。高精度的回转工作台由带间隙补偿的双导程蜗杆驱动。

图 5.10　手动操作的回转工作台

回转工作台用于下列场合。

（1）在六角螺栓上铣六面（扳手用）。
（2）在圆法兰上钻均匀分布的孔。
（3）凸柄上加工圆角。
（4）小型铣床无法驱动大钻头（约＞13mm），通过圆形刀具路径铣削加工完成大孔钻削。
（5）螺旋铣削。
（6）加工复杂的曲线（合适装夹）。
（7）任意角度线性切削。
（8）加工圆弧。
（9）通过在水平工作台上加一个回转工作台组成复合工作台，用户能够将刀具中心移动到被加工零件的任何位置，这样就可以在零件的任何位置加工圆弧。
（10）加工圆弧面。

此外，如果改为采用步进电动机控制，CNC 铣床上的回转工作台配合动力架能够完成需要车床加工的零件。

5.3 CNC 机床的刀具系统

自动换刀装置如图 5.11 所示，CNC 机床用于下列场合。
（1）普通机床难以加工的形状复杂的零件加工。
（2）小批生产，如单件（一次性）生产、原型制作、工具制造等。
（3）加工精度要求高和重复加工的零件加工。
（4）需要多次装夹且装夹成本高的零件加工。
（5）需要多次更改设计才能定型且制造成本高的零件加工。
（6）难测量且测量成本高的零件加工。

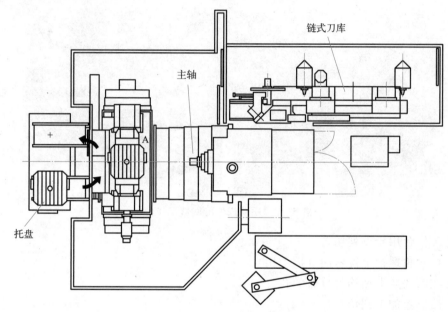

图 5.11 自动换刀装置

5.3.1 刀具

1. 刀具材料

常用的刀具材料见表 5-1。

表 5-1 常用的刀具材料

刀具材料	工件材料	备注
碳素工具钢	低强度钢、韧性材料、有色合金、塑料	低切削速度、低强度材料
低合金钢、中合金钢	低强度钢、韧性材料、有色合金、塑料	低切削速度、低强度材料
高速钢	中低强度和中低硬度的材料	中低切削速度、中低强度材料

续表

刀具材料	工件材料	备注
硬质合金	中等强度和中等硬度的材料	不适合低速加工
涂层硬质合金	铸铁、合金钢、不锈钢、超级合金	不适用于钛合金，也不适用于有色合金，因为此时涂层跟非涂层区别不大
陶瓷	铸铁、镍基超级合金、有色合金、塑料	不适用于低速加工或断续切削加工，不适用于加工铝合金、钛合金

（1）碳素工具钢。

（2）低合金钢、中合金钢。

（3）高速钢。

（4）硬质合金。以下建议将有助于选择硬质合金刀具。

① 钴含量越低、粒度越小、强度越高的硬质合金，其切削质量越好。

② 如果加工材料（钢除外）没有坑凹、磨损或擦伤，则使用含碳化钨（WC）的硬质合金。

③ 为了减少钢加工时的坑凹和磨损，使用含碳化钛（TiC）的硬质合金。

④ 对于高温、高压、韧性大的钢材的强力切削，使用含有钨-钛-钽（W-Ti-Ta）或低黏合剂的复式碳化物硬质合金。

（5）涂层硬质合金。

（6）陶瓷。以下建议将有助于选择陶瓷刀具。

① 推荐使用高速切削，选择具有大刀尖半径的方形刀片或圆形刀片。

② 使用主轴转速高、自锁夹紧、刚性好的机床。

③ 工件刚性好。

④ 确保电源稳定且不间断（避免中途停车）。

⑤ 使用负前角，以便减少直接施加到陶瓷刀具刀尖的应力。

⑥ 刀架悬伸尽可能短，一般不超过刀柄厚度的1.5倍。

⑦ 陶瓷刀片采样大刀尖半径和侧刃切削角，以减少崩刃。

⑧ 采用大切深（背吃刀量）、低进给，而不是小切深、高进给加工，陶瓷刀片的切削深度能够达到刀片切削面宽度的一半。

⑨ 对氧化铝基陶瓷刀具，避免使用冷却剂。

⑩ 换用陶瓷刀具时，重新安排加工顺序；如有可能，则引入端倒角或降低进给速度。

2. 刀具应用

刀具是通过剪切变形方式从工件上去除材料的切削工具。切削可通过单刃刀具或多刃刀具完成。单刃刀具用于车削、成形、刨削等类似加工，通过一个切削刃去除材料。铣削和钻孔的刀具通常是多刃工具。研磨工具也是多刃刀具，每一粒磨料就像一个微观的单刃刀具（具有高的负前角），每次剪切一个微小的碎片。刀具结构如图5.12所示。

刀具材料的硬度要比待加工工件的硬度高，并且刀具能够承受金属切削过程中产生的切削热。此外，刀具必须具有特定的几何结构，后角的设计应使切削刃能够接触工件，而

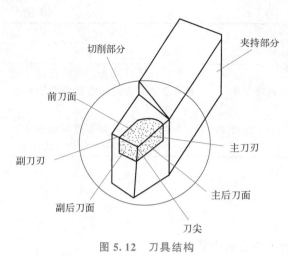

图 5.12 刀具结构

刀具的其余部分无须接触工件。切削面的角度、凹槽（如断屑槽）宽度、凹槽数或齿数及棱边尺寸是刀具设计时必须考虑的。为了提高刀具的使用寿命，切削速度和进给量等因素都必须优化。

硬质合金刀片的形状如图 5.13 所示。

（1）V（35°菱形）——用于轮廓精加工，强度最低，每侧有两个切削刃。

（2）D（55°菱形）——强度稍高，在角度允许时也可用于轮廓精加工，每侧有两个切削刃。

（3）T（三角形）——每侧有三个切削刃，常用于车削。

（4）C（80°菱形）——最常用的硬质合金刀片，用一个刀架完成转位对刀，每侧有两个切削刃。

（5）W（80°三角形）——新型号，可以像 C（80°菱形）一样转位对刀，每侧有三个切削刃。

（6）S（方形）——强度高，主要用于倒角，不适合加工台阶面，每侧有四个切削刃。

（7）R（圆形）——强度最高，但不常用。

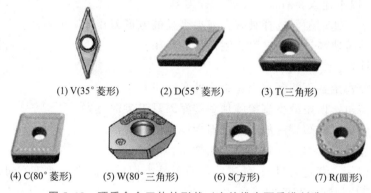

图 5.13 硬质合金刀片的形状（山特维克可乐满制造）

车削（图 5.14）即车床加工，如刀具从工件的外圆上去除金属材料。

螺纹车削（图 5.15）是在底孔中加工出内螺纹的过程，以便螺钉或螺栓能够旋入内螺纹。它还用于在螺母上加工螺纹。

立式铣床使用旋转刀具加工零件的平面等。立式铣床属于轻型机床，所用的铣刀相对灵活，适用范围较广。铣削如图 5.16 所示。

镗孔是指用钻头在实体材料上扩大或精加工孔的操作，可以在车床或铣床上完成。

研磨是一种通过使用磨料颗粒进行精整加工的操作。研磨过程通过大量小的单颗磨粒的相对运动，可以大量去除非常小的碎片。

钻削是用钻头去除大量金属以获得半精密孔或型腔的经济加工方法，如图 5.17 所示。

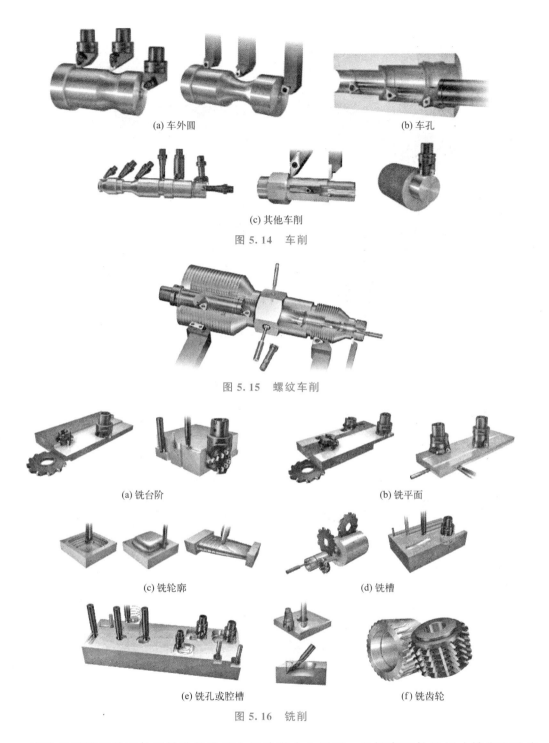

(a) 车外圆　　　　　　　　　　　　(b) 车孔

(c) 其他车削

图 5.14　车削

图 5.15　螺纹车削

(a) 铣台阶　　　　　　　　　　　　(b) 铣平面

(c) 铣轮廓　　　　　　　　　　　　(d) 铣槽

(e) 铣孔或腔槽　　　　　　　　　　(f) 铣齿轮

图 5.16　铣削

　　铰孔是用铰刀从工件已钻孔的孔壁上切除微小金属层，以提高工件的尺寸精度和孔表面质量的方法。

　　珩磨是一种孔加工技术，它利用旋转珩磨头上的珩磨条（也称油石）对精加工表面进行精整加工，这些孔都要求很高的表面粗糙度。

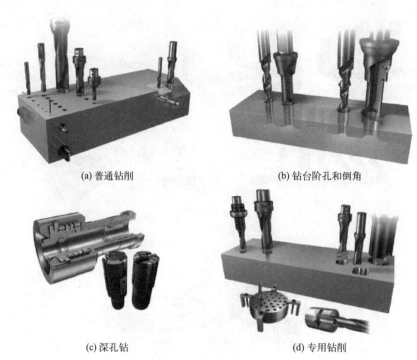

(a) 普通钻削　　　　(b) 钻台阶孔和倒角

(c) 深孔钻　　　　(d) 专用钻削

图 5.17　钻削

3. 刀具装夹机构

刀具装夹机构的应用如图 5.18 所示。

(a) 车削　　　　(b) 铣削

(c) 钻削　　(d) 镗削　　(e) 螺纹车削

图 5.18　刀具装夹机构的应用

典型的刀具装夹机构如图 5.19、图 5.20 和图 5.21 所示。

4. 自动测量系统

自动测量系统（图 5.22）在 CNC 机床中的应用如下。

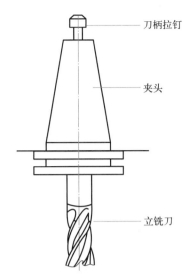

图 5.19 主轴刀具（立铣刀）装夹机构

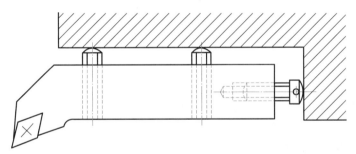

图 5.20 车刀装夹机构

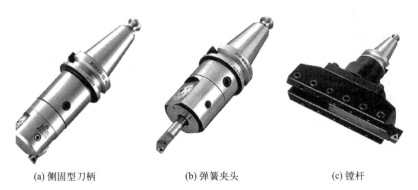

(a) 侧固型刀柄　　　　(b) 弹簧夹头　　　　(c) 镗杆

图 5.21 铣刀装夹机构

(1) 找工件基准。
(2) 工件尺寸测量。
(3) 刀具偏置测量。
(4) 刀具破损监测。
(5) 数字化控制。

自动测量系统在 CNC 机床中的应用示例如图 5.23 所示。

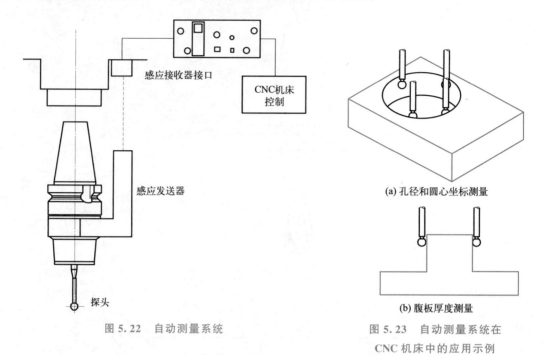

图 5.22　自动测量系统

图 5.23　自动测量系统在 CNC 机床中的应用示例

5.3.2　自动换刀装置

1. 常见的自动换刀装置

CNC 机床配有自动换刀装置（automatic tool changer，ATC）。根据机床类型和应用场合，ATC 可以同时取下不同数量的刀具，并把数控程序要调用的刀具装到机床主轴上。最常见的 ATC 有转塔回转刀架和刀库。

转塔回转刀架（图 5.24）主要用于 CNC 车床中，刀库主要用于 CNC 铣床中。数控程序发出换刀信号，转塔旋转，所需刀具到达工作位置。整个换刀过程只需数秒。

根据机床类型和尺寸的不同，CNC 车床的转塔回转刀架设置有 8～16 个刀位。在大型加工中心，最多可同时使用 3 个转塔回转刀架。如果使用的刀具超过 48 把，加工中心中需要装配不同类型的刀库，刀库容量可达 100 把刀具，甚至更多。刀库有径向刀库、环形刀库、圆盘式刀库、链式刀库（图 5.25）和盒式刀库等。

在刀库中，使用刀具夹持器（换刀装置）进行换刀，如图 5.26 所示。在 CNC 系统发出换刀指令后，双爪夹紧装置进行换刀，过程如下。

(1) 将所需刀具移动到换刀位置。
(2) 将主轴置于换刀位置。
(3) 旋转换刀装置，双爪分别抓住主轴与刀库中的刀具。
(4) 将刀具拉出，将换刀装置旋转 180°。
(5) 换刀装置到位后，将刀具分别放到主轴和刀库中。

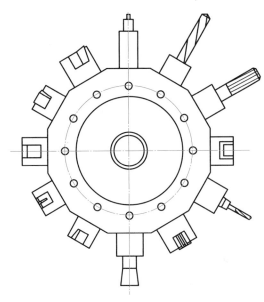

图 5.24 转塔回转刀架

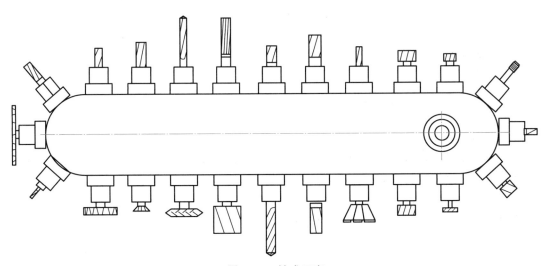

图 5.25 链式刀库

(6) 将换刀装置复位,完成换刀。

换刀过程一般需要 6~15s,最快能在 1s 内完成换刀。

2. 自动换刀装置

自动换刀装置实现自动换刀需要以下条件支持。

(1) 能够储存足够数量刀具的刀具库。
(2) 方便换刀机械手取刀的刀具传送装置。
(3) 控制系统能够实现换刀功能。
(4) 换刀程序。

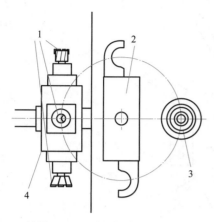

1—铣削刀具；2—刀具夹持器（换刀装置）；
3—主轴；4—刀库
图 5.26　自动换刀装置

机械手换刀动作的具体流程如下。

(1) 在机械手的合适方向停止主轴，以便从主轴上换下刀具。

(2) 将机械手移动到主轴换刀位置。

(3) 机械手从主轴上取刀。

(4) 机械手到达刀库。

(5) 刀库给出从主轴换下的刀具指定位置。

(6) 将刀具放入刀库中。

(7) 对刀库进行检索，将所需刀具移动到换刀位置。

(8) 机械手从刀库中选取刀具。

(9) 机械手到达主轴换刀位置。

(10) 新刀具放入主轴中。

(11) 机械手返回原始位置。

CNC 机床的安全规范。

安全工作的目的是消除对现场工作人员、机床和设施造成损害的事故。

一般情况下，CNC 机床的工作安全预防措施与传统机床相同，可以分为以下三类。

(1) 消除危险。

① 机床和工作相关的所有设备问题需要及时登记。

② 急停退出必须有效。

③ 禁止在衣物中携带尖锐物体。

④ 现场工作人员不能佩戴手表和戒指。

(2) 防护和标记危险区域。

① 不允许删除或停用安全防范措施和相应的警示标志。

② 必须对移动和交错部位进行防护。

(3) 消除暴露的危险。

① 现场工作人员必须穿防护服，以防可能的火花和闪光出现。

② 现场工作人员必须佩戴防护眼镜或防护罩以保护眼睛。

③ 禁止使用损坏的电缆。

本 章 小 结

通过本章学习，要求学生了解 CNC 机床的机械系统，尤其是主轴、导轨、工作台的结构，以及刀具系统；理解 CNC 机床的选用和故障诊断；掌握 CNC 机床刀具选用和故障诊断方法；通过对自动换刀装置等的学习能够对 CNC 机床自动上下料、生产线自动化进行改造升级；掌握数控技术常用词汇，如 CNC lathe、tool magazine、high speed steel、tungsten carbide、ceramic、chuck、vice、headstock、carriage、turret、tailstock、machining center、driving lead screw、chip conveyor、spindle、ball screw、coolant through the spindle、guide rail、rolling guide、sliding guide、rotated working table、turning、tapping、grooving、milling、drilling、reaming、honing、end mill、automatic tool changer 等。

本章的知识和能力图谱如图 5.27 所示。

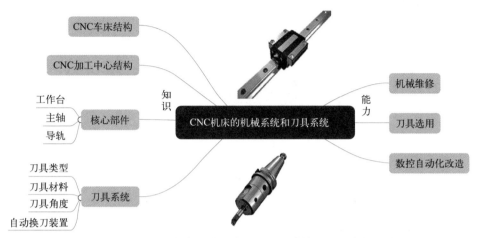

图 5.27　CNC 机床的机械系统和刀具系统的知识和能力图谱

第 6 章

CNC 技术发展

教学目的及要求

了解开放式 CNC 系统；

了解 STEP-NC；

了解数控技术的高级应用；

了解智能制造技术。

6.1 开放式 CNC 系统

20 世纪 50 年代初，麻省理工学院成功开发出 CNC 系统，依赖于微处理器的出现和发展，CNC 系统得以快速发展。20 世纪 70 年代自动化系统的出现，推动了 CNC 系统的跃变。鉴于 CNC 系统的复杂性，它不仅包括普通的控制功能，而且集成了各种辅助功能，如加工方案生成、工艺路线设计及智能化技术。CNC 系统市场主要由日本和德国的企业垄断。为了防止技术外泄或失控，维持市场份额，高档数控产品厂家把 CNC 系统设计成封闭结构。20 世纪 80 年代中期以后，随着计算机技术的发展，计算机网络和优化技术不断应用到制造系统中，其与高速、高精加工控制功能的生产要求一起推进了 CNC 系统向新的模式转变。

传统的 CNC 系统是一种专用的封闭式系统，它越来越不能满足制造模式的发展，难以满足用户特殊的需要，即产品的个性化要求。CNC 系统的改进依赖于 CNC 系统开发商，而不是 CNC 机床制造商，仅靠 CNC 机床制造商的有限资源难以推动 CNC 系统的变革。

开放式 CNC 系统是 CNC 技术研究的热门，作为开放式体系结构研究成果，PC-NC 于 20 世纪 90 年代初被引入，与 20 世纪 80 年代初出现的 IBM-PC 技术一样，在第三方开放式技术推动下，CNC 系统向基于 PC 技术的开放式 PC-NC 发展。虽然 PC-NC 成本低、开放性强、开发产品丰富，但 NC 部分仍然很传统，应用软件的可靠性和开放性不够，用户难以介入 CNC 系统的核心。开放式 CNC 系统的发展历程如图 6.1 所示。

图 6.1 开放式 CNC 系统的发展历程

开放式 CNC 系统应具备以下特点。

(1) 互操作性。互操作性是指构成系统内部的各部件一起协调工作的能力。为此，要求各组件具有标准协议的数据格式、行为模型、通信接口、通信方式和交互机制，如基于总线的系统开发。

(2) 可移植性。可移植性是指一个组件可以不加改动地从一种运行环境转移到另一种运行环境下运行。从商业角度来看，可移植性非常重要，即 CNC 硬件或软件功能模块能运行于不同供货商提供的平台上，这大大提高了平台的工作效率。

(3) 可扩展性。可扩展性是指系统功能模块化，使实时增减功能模块经济可行。典型的例子是添加内存条或加板卡。

(4) 可互换性。可互换性是指具有统一的接口，相同功能的模块可以相互替换。整个系统不需要推倒重来。典型的例子是用新算法运动控制卡替换旧卡。

作为开放式 CNC 系统，要考虑模块化、可扩展、可重用、兼容性好，这些反映在上述的开放式 CNC 系统特点中。换个角度看，开放式 CNC 系统可以看作一个柔性化标准系统，柔性化与互操作性和可扩展性含义相似，标准化反映可移植性和可互换性。

开放式 CNC 系统可作为中、小型 CNC 机床制造商升级换代的一种途径。从最近的研究成果来看，开放式 CNC 系统由于其柔性化、生产周期短和质量可靠而受到 CNC 机床制造商青睐。具有代表性的专注于 CNC 机床 PC 基础控制的系统是 OAPC-NC 系统，OAPC-NC 系统实现了软硬件集成（图 6.2）。

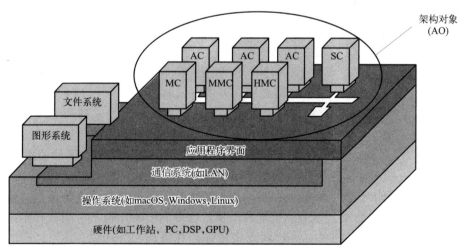

图 6.2 OAPC-NC 系统

开放式 CNC 系统日新月异，高速通信促进了不同类型的开放式 CNC 系统的发展。这些开放式 CNC 系统大多将普通 PC 的开放性与传统数控功能相结合，即"CNC＋PC" CNC 系统。开放式 CNC 系统的优势在于：即使在机器硬件老化的情况下，也可以使 CNC 功能保持最佳技术状态，满足加工过程需求。例如，某个模具加工 CNC 系统，可以把第三方软件添加到其开放式 CNC 系统中，增强已有功能或增加新功能。在开放式 CNC 系统中，最常见的功能如下。

（1）低成本的网络通信技术。

（2）以太网。

（3）自适应控制。

（4）条码读取器接口、刀具 ID 读取接口、托盘 ID 系统接口。

（5）海量数控程序存储和编辑。

（6）统计过程控制（statistical process control，SPC）数据采集。

（7）文档处理。

（8）CAD/CAM 集成或车间现场编程。

（9）通用的 CNC 人机界面。

用户比较关心开放式 CNC 系统的易用性，这种易用性主要指从一个 CNC 系统到另一个 CNC 系统操作的通用性。一般情况下，由于 CNC 机床类型的差异和 CNC 机床制造商不同，CNC 人机界面不同，操作员必须分别接受不同 CNC 机床或制造商的培训。开放式 CNC 系统可以跳过不同制造商，开发通用的 CNC 人机界面。

现在，CNC 机床用户可以为数控操作创建自己的用户界面，而不必像程序员一样从头开发。此外，开放式 CNC 系统具有用户管理功能，可对不同人员（操作员、程序员、维护人员等）分别授权，登录后只能看到在其授权范围内的功能，其他功能不再显示，这样使数控操作更加简单。

6.2 STEP-NC

随着信息技术与数控技术的快速进步，近十年以来制造环境已经发生了显著的变化。然而，作为 CAM 与 CNC 之间接口的底层标准——G 代码和 M 代码已有 70 多年的历史，但目前其局限性已经显现，成为全球协同智能制造的障碍。一种面向对象的新型 NC 编程数据接口国际标准 STEP-NC 正在世界范围内开发，以取代 G 代码和 M 代码。

STEP-NC 将涵盖数字化制造的整个过程，正逐步取代铣削、车削和电火花加工等旧的 G 代码和 M 代码，并在制造企业推广。目前，STEP-NC 数据传输模型已经建立，STEP 兼容式 CAD/CAM/CNC 系统集成开发与实现正日益受到世界各国的关注。

如图 6.3(a) 所示，G 代码包含主轴速度、指令 G 代码、坐标轴移动值、刀具功能、进给速度和辅助功能。仅有这些信息，机床操作者很难通过数控程序弄懂操作流程、加工条件和刀具性能。此外，数控装置不可能自动执行和智能控制，也不可能在有限的信息下处理紧急情况。相比之下，STEP-NC 包含了所需的任务信息 [图 6.3(b)]，如工步、切削加工、加工刀具、工艺、机床功能、加工策略和制造特征。换言之，STEP-NC 包含了更丰富的信息集合，解决了"加工什么"（几何信息）和"如何加工"（工艺规划）的问题。

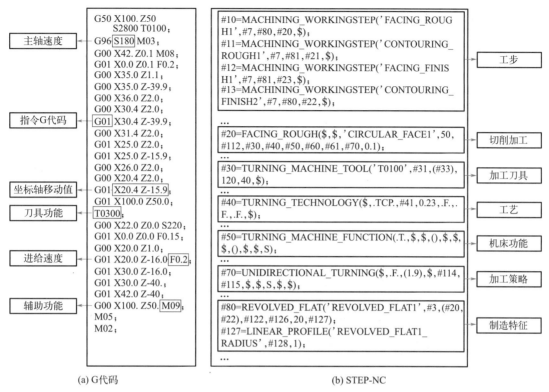

图 6.3　G 代码和 STEP-NC 的对比

6.3　数控技术的高级应用

6.3.1　柔性制造系统

柔性制造单元（flexible manufacturing cell，FMC）可看作柔性制造系统（flexible manufacturing subsystem，FMS）的子系统。FMC 与 FMS 之间存在以下差异。

（1）FMC 不受中央计算机的直接控制，中央计算机发出的指令被传送到其单元控制器。

（2）FMC 能加工的零件种类数量有限。

通常 FMC 由单元控制器、PLC、数台 CNC 机床、物料搬运设备（工业机器人或托盘）等组成。

FMC 按顺序对零件流执行固定的加工操作。

FMS 包括两个子系统：物理子系统和控制子系统。

FMS 的物理子系统由以下部分组成。

（1）加工中心。它由 CNC 机床、普通机床、检测设备、装卸处理设备和加工区等组成。

（2）自动化立体仓库。它在在制品期间充当缓冲区，并保存在加工或操作之间临时存储零件的设备，如传送带。

（3）物料搬运系统。它由小车、传送带、自动导引车等组成，用于在加工工作站之间搬运零件。

FMS 的控制子系统由以下部分组成。

（1）控制硬件。它由微型计算机、PLC、通信网络、交换设备、打印机、大容量存储器等组成，以提高 FMS 的加工能力。

（2）控制软件。它是一组用于控制物理子系统的文件和程序。

数控程序在 FMS 编程模块中生成并可进行仿真，如 SL FMS（图 6.4）通过控制模块在机床上作进一步处理，经由键盘/鼠标的交互操作完成修改，如用标准 PC 键盘或其他特定键盘来比较新旧数控程序。数控程序的修改在相应编程系统编辑器中完成。SL FMS 的组成及应用领域如图 6.5 所示。

图 6.4　SL FMS

FMS 物理子系统各组成部分的基本特征如下。

（1）CNC 机床。

CNC 机床是 FMS 的主要组成部分，因为它决定了 FMS 的灵活性和加工能力。CNC 机床的性能要求如下。

① 大多数 FMS 使用卧式加工中心或立式加工中心。立式加工中心比卧式加工中心适应性差些（由于立柱和刀具更换装置的高度限制）。

② 加工中心对所有方向的运动进行控制，如主轴在 X、Y、Z 方向上的运动，工作台的回转、摆动，以保证加工的高灵活性。

③ 加工中心能够执行多种操作，如车削、钻孔、轮廓加工等。通过托盘物料搬运接口，集成托盘交换器、自动存储和检索系统，实现加工中心内部和加工中心之间的物流搬运。

（2）夹具和刀库。

夹具和刀库具有以下特点。

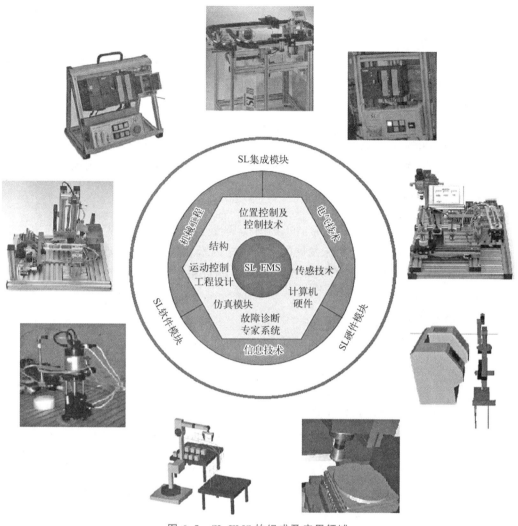

图 6.5 SL FMS 的组成及应用领域

① 在数控加工中,首先要把加工工件装夹在 CNC 机床上的确定位置,并保证其装夹的定位精度,使其装卸方便、生产准备时间短。组合夹具作为 CNC 机床的一种快速装夹工具,可以节省生产费用。夹具广泛应用于自动化立体仓库和物料搬运系统(如自动导引车)中。

② 所有加工中心都配有刀库,即刀具存储系统。刀库可以索引最常用的刀具,以确保非加工时间最优。此外,使用快速换刀装置、刀具适时重磨和提供充足备件有助于减少非加工时间。

(3)物料搬运设备。

在 FMS 中使用的物料搬运设备包括工业机器人、传送带、无人搬运车、单轨和其他轨道车辆、特种车辆等。其特点如下。

① 与加工中心、自动化立体仓库集成。

② 对于形状不规则的零件,物料搬运系统配有模块化托盘夹具。对于回转零件,用

工业机器人实现CNC车床的零件装卸和加工中心之间的输送。

③ 搬运系统必须由计算机系统直接控制，以便导入各加工中心、装卸点和储存区。

（4）检验设备。

检验设备包括用于离线检测的三坐标测量机，通过编程测量零件表面的尺寸、同轴度、垂直度和平面度等。检验设备的特点是与加工中心可以无缝连接。

（5）其他组成部分。

其他组成部分包括中央冷却系统和高效切屑分离系统。其特点如下：

① 必须有冷却液回收功能。

② 加工和检查前必须对工件、夹具和托盘等进行适当清洗，以清除污垢和切屑。

6.3.2 CNC机床在模具制造中的应用

数控技术日新月异，这些变化有助于提高模具行业数控加工的生产效率。其中，运算速度更快的CPU是许多CNC机床发展的主要影响因素。数控技术的改进不仅体现在加工速度的提高上，还涉及许多CNC关联技术的进步。近年来，随着数控技术的发展，模具制造中数控加工越来越多。

1. 程序段处理时间及其应用

随着CPU运算速度的提高和CNC机床制造商把高性能CPU融入集成化的CNC系统，CNC性能显著提高。虽然反应更快、更灵敏的CNC系统可以实现更高的程序处理速度，但事实上，它也会存在一些潜在的问题，这些问题有可能成为限制加工速度的瓶颈。

如今，大多数模具车间都清楚地认识到，高速加工不仅靠减少程序段处理时间就能实现。在许多方面，以赛车为例，更能够说明原因。最快的赛车一定能够赢得比赛吗？即使是一个偶尔才观看车赛的观众都知道，除速度外，还有许多因素会影响比赛的结果。首先，车手对于赛道的了解程度很重要，其必须知道何处有急转弯，以便能恰如其分地减速，从而安全高效地通过弯道。在采用高进给速度加工模具的过程中，CNC系统中的待加工轨迹监控技术可预先获取锐角曲线出现的信息，这一功能起着同样的作用。

车手对其他车手动作及不可确定因素的反应灵敏程度与CNC系统中伺服反馈的反应时间类似。CNC系统中伺服反馈主要包括位置反馈、速度反馈和电流反馈。

当车手驾车绕赛道行驶时，动作的连贯性，能否熟练地制动、加速等对车手的临场表现有着非常重要的影响。同样，CNC系统的钟形加速、减速和待加工轨迹监控功能利用平稳加速、减速来代替突然变速，以保证机床加减速的平稳性。

除此以外，赛车和CNC系统还有其他相似的地方。赛车发动机的功率类似于CNC系统的驱动装置和电动机；赛车的质量可以和CNC机床中运动构件的质量相提并论，赛车的刚度和强度则类似于CNC机床的强度和刚度；CNC系统修正指定刀具轨迹误差的能力与车手具备的将赛车控制在车道内的能力极其相似。

另一个与CNC机床相似的例子是：那些速度不是最快的赛车往往不需要技术全面的车手（最快的赛车更需要技术全面的车手）。过去只有高档的CNC机床才能在高速切削的同时保证较高的加工精度。如今，中、低档的CNC机床所具备的功能也有可能令人满意地完成工作。虽然高档CNC机床具备目前所能获得的最佳性能，但也存在这种可能，即用低档 CNC 机床获得同类产品中、高档 CNC 机床的加工精度。过去，限制模具制造最高进给速度的因素是 CNC 系统，如今则是 CNC 机床的机械结构。在 CNC 机床已处于性能极限的情况下，更先进的 CNC 系统也难以提高其性能。

2. CNC 系统的基本特性

以下是目前模具制造过程中 CNC 系统的基本特性。

(1) 曲线、曲面的非均匀有理样条（NURBS）插补。

该项技术采用沿曲线插补的方式，而不是用一系列短直线来拟合曲线。这一技术的应用已经相当普遍，许多模具制造行业目前使用的 CAM 软件都提供了该选项，即生成 NURBS 插补格式的数控程序。同时，功能强大的 CNC 系统提供了五轴插补功能及其相关的特性。这些特性提高了零件表面加工质量，改善了电动机运行的平稳性，提高了切削速度，并使数控程序更简捷。

(2) 更小的指令单位。

大多数 CNC 系统向机床主轴传递运动和定位的指令单位都在 $1\mu m$ 以上。随着 CPU 处理能力的不断提高，一些 CNC 系统的最小指令单位已经达到 1nm（0.000001mm），可获得更高的加工精度，电动机运行也更加平稳。电动机运行的平稳性使一些 CNC 机床能够在床身振动不加大的前提下提高切削速度。

(3) 钟形曲线加减速控制。

钟形曲线加减速控制又称 S 曲线加减速控制或冲击控制。与使用直线加减速控制相比，这种控制方式可使 CNC 机床获得更好的加减速控制效果。与直线加减速控制和指数加减速控制等加减速控制相比，采用钟形曲线加减速控制可获得更小的定位误差。

(4) 待加工轨迹监控。

这一技术已被广泛使用，其性能千差万别，是高档 CNC 系统有别于低档 CNC 系统的主要指标。总的来讲，CNC 系统就是通过加工轨迹监控来实现对程序的预处理，以此来确保能获得更优异的加减速控制。根据不同的 CNC 系统的性能，待加工轨迹监控所需的程序段数量从两个到上百个不等，这主要取决于数控程序的最短加工时间和加减速控制的时间常数。一般而言，要想满足加工要求，就要有一个待加工轨迹监控程序段最小值（如15 个）。

(5) 数字伺服控制系统。

数字伺服控制系统发展迅速，以至于大多数 CNC 机床制造商都选择数字伺服控制系统。在该系统中，快速通信、驱动器与 CNC 系统的串行连接和高速多核数字信号处理器的集成使 CNC 系统对机床的控制更加紧凑、精确。数字伺服控制系统的作用如下。

① 提高电流回路的采样速度，再加上电流控制的改善，可以降低电动机温升。这样，不仅可以延长电动机的使用寿命，还可以减少传递到滚珠丝杠的热量，从而提高滚珠丝杠的精度。除此之外，采样速度的加快可以提高速度闭环增益，这些都有助于提高 CNC 机床的整体性能。

② 由于许多新的 CNC 系统使用高速串行接口与伺服控制系统相连，因此通过通信网络，CNC 系统可获得更多的电动机和驱动装置的工作信息，这些可提高机床的可维护性。

③ 连续位置反馈允许在高速进给的情况下进行高精度的加工。CNC 系统运算速度的

加快使位置反馈速度成为制约机床运行速度的瓶颈。在传统的反馈方式中，反馈速度受到信号类型的制约，如 CNC 系统和外部编码器电子元器件之间的采样速度。若采用串行反馈，则这一问题将得到很好的解决，即使 CNC 机床以很高的速度运行，也可达到精密的反馈精度。

近几年来，直线电动机的工作性能和认可度有了显著的提高，故很多加工中心采用直线电动机。GE Fanuc 开发的一些先进技术使 CNC 机床上的直线电动机的最大输出力达 15500N，最大加速度达 $30g$。其他先进技术的应用使 CNC 机床的尺寸得以减小，质量得以减小，冷却效率大幅提高。所有这些技术上的进步使直线电动机在与旋转电动机相比时，优势更强：更大的加减速率、更准确的定位控制、更高的刚度、更高的可靠性、内部的动态制动。

3. CNC 五轴加工中心

在制造复杂模具的过程中，CNC 五轴加工中心（图 6.6）的使用越来越普及。使用 CNC 五轴加工中心可以减少单个零件加工所需的作业准备次数和机床的数量，将在制品库存减至最低，降低总加工时间。

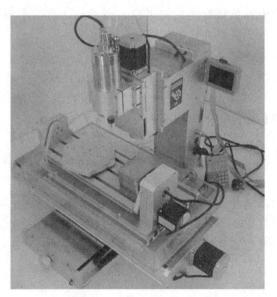

图 6.6　CNC 五轴加工中心

CNC 五轴加工中心的功能越来越强大，过去只有高档 CNC 五轴加工中心才具备的功能，如今也被用在中档 CNC 五轴加工中心上。对于那些从未使用过 CNC 五轴加工技术的加工企业而言，这些特性的应用使 CNC 五轴加工中心变得更简单。将目前的 CNC 技术用于五轴加工中心，CNC 五轴加工中心将具备以下优势。

（1）减少专用工具的需求。
（2）允许在完成数控程序后再设定刀具的偏置。
（3）支持通用程序设计，这样经过后处理的程序可以在不同 CNC 机床之间互换使用。
（4）提高表面精加工的质量。
（5）可用于不同结构的 CNC 机床，这样就不必在程序中说明是主轴转动还是工件绕中枢轴转动，这将由 CNC 的参数来解决。

以球形铣刀的补偿为例来说明 CNC 五轴加工中心为什么特别适合于模具制造。在工件和刀具绕中枢轴转动时，为了准确补偿球形铣刀的偏置，CNC 五轴加工中心必须能够在 X、Y、Z 三个方向动态地调整刀具的补偿量，保证刀具切触点的连续，这有利于提高精加工的精度。

此外，CNC 五轴加工中心的用途表现在：刀具绕中枢轴转动相关的特性调整、工件绕中枢轴转动相关的特性调整及允许操作者采用手动方式改变刀具矢量的特性。

当采用刀具的中枢轴作为回转轴时，原来 Z 轴方向的刀具长度偏置将被分成 X、Y、Z 三个方向的分量。另外，原来 X 轴、Y 轴方向的刀具直径偏置被分为 X、Y、Z 三个方向的分量。由于在切削工程中刀具可以沿旋转轴方向做进给运动，因此这些偏置必须动态更新，以便确定连续变化刀具的方位。

CNC 五轴加工中心还具有刀具中心点编程的特性，允许编程人员定义刀具的路径和中心点速度，CNC 五轴加工中心通过旋转轴和直线轴方向的指令来保证刀具按照程序运动。这一特性使刀具的中心点不再随刀具的变化而变化，这也意味着 CNC 五轴加工中心可以像 CNC 三轴加工中心一样，直接输入刀具的偏置；不必再次说明程序或计算刀具长度的改变。这种通过使主轴旋转来实现转轴运动的特性简化了刀具的编程后置处理。

利用同样的功能使工件绕中枢轴旋转，机床也可以获得旋转运动。新研制的 CNC 五轴加工中心能够通过动态调整固定偏置和旋转坐标轴来配合零件运动。

当操作人员采用手动方式来实现机床的慢速进给时，CNC 系统同样起着重要的作用。新研制的 CNC 五轴加工中心同样允许轴沿刀具向量的方向缓慢进给，在没有刀具位置变化的前提下，还允许改变刀具向量的方向。这些特性使操作人员在使用 CNC 五轴加工中心的过程中，能够很容易地使用目前在模具制造业广泛使用的"3+2 编程法"。然而，随着新的 CNC 五轴加工功能的逐渐发展和这些功能逐渐被接纳，CNC 五轴加工中心将会更加普及。

6.3.3 虚拟轴机床

虚拟轴机床又称六足机床或并联机床。

虚拟轴机床采用 Stewart 平台，属于多轴加工中心，通过控制六条六自由度"腿"的长度就可以控制装有主轴头的动平台在空间的位置和姿态，实现具有六自由度运动的复杂曲面加工。由于虚拟轴机床继承了并联机构的优点，因此其为制造加工提供了更好的性能，即结构刚度高、有效载荷大、运动速度大、能够实现高速和高精度的加工。

虚拟轴机床包含以下六个子系统。

(1) 机床框架。
(2) 带有铣削主轴头的动平台。
(3) 伸缩执行机构。
(4) 自动换刀装置。
(5) 工件自动交换系统。
(6) 带控制系统电源的电气开关柜。

虚拟轴机床的基本结构如图 6.7 所示，它由工件交换位置、带铣削主轴头的动平台、电气开关柜等组成，其主要特征是在整个机床中间有一个并联机构（图 6.8）。

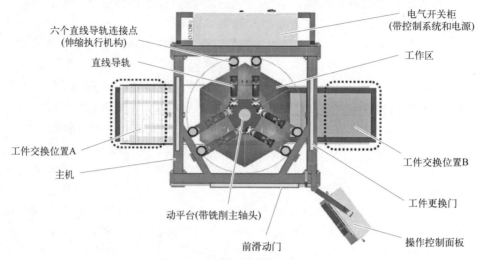

图 6.7　虚拟轴机床的基本结构

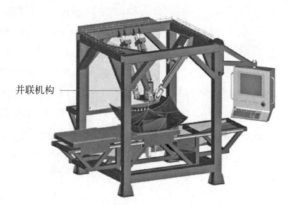

图 6.8　虚拟轴机床的并联机构

6.3.4　虚拟计算机数控系统

虚拟计算机数控（virtual CNC，VCNC）系统可以在设计阶段预测和优化 CNC 机床的动态性能。通过在 VCNC 系统上运行零件数控程序并对加工正确性进行评估，可以在机床生产之前评估导轨、驱动器、编码器、控制原理和插补算法选择等各种设计选择对数控加工性能的影响。

VCNC 系统还可用于调整伺服控制系统和插值的参数，而不占用实际加工时间。一旦在 VCNC 系统中达到所需的响应速度和轮廓精度，参数就可以在实际 CNC 机床上以最小的停机时间直接实现。在工艺规划中，VCNC 系统可以用来评估不同数控程序的轮廓误差，从而对进给速度和刀具轨迹进行必要的修整，以避免伺服控制系统误差导致加工误差。VCNC 系统的仿真精度依赖于各部件动态行为数学模型的准确性。

利用 VCNC 系统的精确仿真能力，对实际 CNC 机床的动态性能进行预测和完善，现在已经开发出以下应用。

(1) 数控编程轮廓误差的预测。
(2) 进给驱动伺服控制器的自动调整。
(3) 样条插值的锐角跟踪。
(4) 虚拟驱动器模型的快速识别。

VCNC 系统的结构如图 6.9 所示，与开放式 CNC 系统结构相似。VCNC 系统接受 CAD/CAM 系统中 CL 文件生成的刀具路径规划，CL 文件被译码为实现加工的刀具运动，包括直线、圆弧和样条曲线等插补线段。各轴的运动轨迹命令是以刀具路径命令为基础，加上进给类型生成的。进给类型可配置为分段常数、梯形或三维加速度突变等。

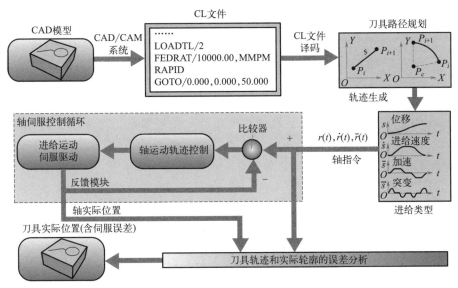

图 6.9 VCNC 系统的结构

通过配置运动控制、进给驱动和反馈模块形成轴伺服循环的闭环控制。运动控制规律从常用控制规律（如 P 控制、PI 控制、PID 控制、P-PI 串级控制和超前滞后控制）中选择，也可以从参考资料中更先进的控制算法（如极点配置、广义预测、自适应滑模控制、前馈控制及摩擦力补偿）中选择。

进给驱动模块可以配置为直接驱动或齿轮驱动的动力学模型。放大器、电动机、惯量轴、摩擦和驱动机构的特性可以完全由数学模型描述，包括非线性效应，如量化、电流和电压饱和、黏滑摩擦和轴向间隙等；也可以包括实验确定或分析预测的具有结构共振的高阶驱动模型。反馈模块可以由线性或角度位置传感器、速度传感器和加速度传感器组成，每个传感器都具有选定的精度和噪声特性。VCNC 系统可以通过运行各部件程序、伺服反馈和轮廓误差，或者通过轴速度、加速度和突变类型及电动机转矩和功率历史数据来评估其性能。通过频域和时域分析，用户可以评估和改善 VCNC 系统轴的稳定裕度和伺服性能。

6.3.5 3D 打印技术

3D 打印机（图 6.10）虽然不能制造我们所需的任何物体，但 3D 打印技术的盛行使越来越多的人坚信，3D 打印机已开始颠覆制造业。

图 6.10　3D 打印机

3D 打印的工作原理是按照计算机的数字指令，使用塑料、陶瓷和金属等打印材料，一层层叠加起来，最终把计算机上的蓝图变成实物。例如，某些 3D 打印机会喷出一股加热的半液体状塑料，随着喷嘴来回移动，打印出物体每一层的轮廓，最后凝固成形。3D 打印样品如图 6.11 所示。

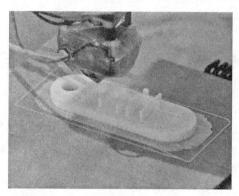

图 6.11　3D 打印样品

3D 打印机采用的指令通常以 CAD 文档的形式存在，这意味着人们可以采用 3D 建模软件在计算机上设计物件，将计算机与 3D 打印机连接，然后 3D 打印机就能将所设计的物件呈现在眼前。3D 打印小飞机如图 6.12 所示。

图 6.12　3D 打印小飞机

1. 3D 打印技术的历史

3D 打印技术也称增材制造技术，已经有 30 多年的历史。1986 年，查尔斯·赫尔发明了第一台商用 3D 打印机，它采用立体光刻技术，在激光的照射下，液态光敏树脂材料固化。

直到 21 世纪初，3D 打印技术才获得更广泛的应用。

新一轮的创业浪潮使 3D 打印在"DIY"领域中流行起来，创客运动就是"DIY"的革新版。3D 打印技术为创客们提供了前所未有的制造能力。

2. 3D 打印技术的未来

3D 打印技术不会取代现代制造业的流水作业方式，其优势在于可以按照需要定制个性化产品，如制造军用飞机的部分专用零件。我国 C919 大飞机装载了 28 件 3D 打印的钛金属舱门件。

医疗行业也利用了 3D 打印技术的这一优势，制造用传统方法很难制造的特殊物件。例如，研究人员建立了 3D 打印的耳朵模型，作为生物工程耳朵的活体细胞框架。

3D 打印技术的普及还大大缩短了买卖双方的距离。通过将打印用的 3D 设计图上传到网上销售，卖方不用支付运费，只需安排所售产品在离买家最近的 3D 打印机上打印即可。

6.4 智能制造技术

6.4.1 人工智能与智能制造

1. 人工智能

人工智能（artificial intelligence，AI）也称机器智能。相对人类的天生智能而言，AI 是指由人制造出来的机器所表现出来的智能。在计算机科学领域，AI 的研究是智能主体的研究与设计，智能主体是指一个可以观察周围环境并作出行动以达到目的的系统。通俗地说，AI 是在机器模仿人类与其他人类思维相关的认知功能时使用的，如"机器学习"（图 6.13）和"问题解决"。

AI 研究的传统问题（或目标）包括推理、知识表示、规划、学习、自然语言处理、感知、移动和操作物体的能力等。通用智能是该领域的长期目标之一。研究方法包括统计方法、计算智能和传统的符号人工智能。AI 中使用了许多工具，包括搜索和数学优化版本、人工神经网络，以及基于统计、概率和经济学的方法。从事 AI 工作的人必须懂计算机科学、数学、心理学、语言学、哲学等诸多学科。

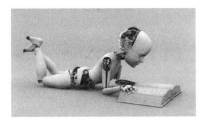

图 6.13 机器学习

现在，AI 发展进入新阶段。经过六十多年的演进，特别是在移动互联网、超级计算、大数据、传感网、脑科学等新理论、新技术及经济社会发展强烈需求的共同驱动下，AI 发展迅速，呈现出深度学习、跨界融合、人机协同、群智开放、自主操控等新特征。大数据驱动知识学习、跨媒体协同处理、人机协同增强智能、群体集成智能、自主智能系统成为 AI 的发展重点，受脑科学研究成果启发的类脑智能蓄势待发，芯片化、硬件化、平台化趋势更加明显，AI 发展进入新阶段。当前，新一代 AI 相关学科发展、理论建模、技术创新、软硬件升级等整体推进，正在引发链式突破，推动经济社会各领域从数字化、网络化向智能化加速跃升。

AI 的应用包括语言识别、策略性游戏（如 AlphaGo）、无人驾驶、智能规划、军事演习、艺术表演（图 6.14）等。

图 6.14　艺术表演

2. 智能制造

智能制造（图 6.15）源于对 AI 的研究，是 AI 在生产过程中的发展和实现。智能制造是指使用先进的技术自动适应不断变化的环境和生产工艺要求，并且能够规划运行生产线和装配线，基本上无须监控和人工干预。

图 6.15　智能制造

智能制造主要有以下研究领域。

（1）机器视觉：二维/三维缺陷检测、智能工业机器人、三维视觉与识别、增强现实显示。

（2）信息物理系统：基于模型的系统工程，如设计、合成和验证"工业 4.0"智能工厂的核心任务。

（3）传感技术：用于制造过程、机器人和环境感知的集成设备模型。

（4）智慧能源、智能电力电子技术：从设计、材料、制造工艺到性能评价、小批量生产等的一站式产品开发与服务。

实现"工业 4.0"的关键在于智能化，我国将从"中国制造"到"中国创造"再到"中国智造"。

3. 智能制造技术

智能制造技术（intelligent manufacturing technology，IMT）是指以计算机仿真为代表的智能技术。智能制造系统（intelligent manufacturing system，IMS）是指基于IMT，借助计算机模拟人类专家的智能系统。

与传统的制造系统相比，IMS具有以下特征。

（1）自组织能力。自组织能力是IMS的一个重要标志。IMS中的各组成单元能够根据工作任务的需要，自行集合成一种柔性的最佳结构，并按照最优的方式运行。其柔性不仅表现在运行方式上，还表现在机构能力上。完成任务后，该结构自动解散，以备在下一个任务中集合成新的结构。

（2）自律能力。IMS具有搜集与理解环境信息及自身的信息，并对信息进行分析判断和规划自身行为的能力。强有力的知识库和基于知识的模型是自律能力的基础。IMS能根据周围环境和自身作业状况的信息进行检测和处理，并根据处理结果自行调整控制策略，以采用最佳运行方案。自律能力使IMS具备抗干扰、自适应和容错等能力。

（3）学习能力与自我维护能力。IMS能以原有的专家知识为基础，在实践中不断学习，充实完善系统的知识库，并删除知识库中不适用的知识，使知识库趋于合理；同时，IMS能对系统故障进行自我诊断、故障排除及修复。这种特征使IMS能够自我优化并适应各种复杂的环境。

（4）智能集成能力。IMS在强调各子系统智能化的同时，更注重整个制造系统的智能集成。这是IMS与制造过程中特定应用的"智能化孤岛"的根本区别。IMS包括各子系统，并把它们集成为一个整体，从而实现系统的智能化。

（5）人机一体化智能系统。IMS不仅仅是AI系统，更是人机一体化智能系统，是一种混合智能。人机一体化突出人在制造系统中的核心地位，同时在智能机器的配合下，更好地发挥人的潜能，使人机之间表现出平等共事、相互理解、相互协作的关系，使两者在不同的层次上各显其能，相辅相成。因此，在IMS中，高素质、高智能的人将发挥更好的作用，机器智能和人的智能将真正地集成在一起，互相配合，相得益彰。

（6）虚拟现实（virtual reality，VR）的应用。虚拟现实是实现虚拟制造的支持技术，也是实现高水平人机一体化的关键技术之一。人机结合的新一代智能界面可用虚拟手段智能地表现现实，它是智能制造的一个显著特征。下面将详细介绍虚拟现实。

6.4.2 虚拟现实

虚拟现实利用计算机生成一种模拟环境，它除了包括听觉和视觉，还包括触觉等感官的模拟，用户如身临其境，模拟环境可以来自现实世界，也可以来自幻想的世界，创造出一种在普通现实世界中不可能的体验。增强现实（augmented reality，AR）也可以被视为虚拟现实的一种形式，它通过摄像机实时将虚拟信息分层传送到头戴式耳机、智能手机或移动设备中，使用户能够查看三维图像。

虚拟现实是仿真技术的一个重要方向，是仿真技术与计算机图形学人机接口技术、多媒体技术、传感技术、网络技术等多种技术的集合，是富有挑战性的交叉技术前沿研究领域。虚拟现实主要包括模拟环境、感知、自然技能和传感设备等方面。模拟环境是由计算机生成的、实时动态的三维立体逼真图像。感知是指理想的虚拟现实应具有一切人所具有的感知。除计算机图形技术所生成的视觉感知外,还有听觉、触觉、力觉、运动等,甚至包括嗅觉和味觉等,也称多感知。自然技能是指人的头部转动、眨眼、手势,或其他人体行为动作,由计算机来处理与参与者的动作相适应的数据,并对用户的输入作出实时响应,分别反馈到用户的五官。传感设备是指三维交互设备。

当前,标准的虚拟现实使用虚拟现实耳机或多投影环境来生成逼真的图像、声音和其他感知,以模拟用户在虚拟环境中的物理存在。使用虚拟现实设备的用户能够环顾人造世界,在人造世界中四处移动,并与虚拟特征或物品互动。这种效果通常是由虚拟现实头戴式受话器创建的,包括头戴式显示器和眼前的显示屏,但也可以通过具有多个显示屏的经过特殊设计的房间来创建。

虚拟现实通过游戏控制器等向用户传输振动和其他感觉的设备称为感知系统。这种感知信息在医疗、游戏、军事训练和驾驶(图 6.16)领域通常被称为力反馈。

图 6.16 虚拟驾驶系统

6.4.3 大数据

数据在快速增长,部分原因是越来越多的廉价信息通过物联网设备(如移动设备、遥感、软件日志、相机、麦克风、电子标签读写器和无线传感网络)被收集。自 20 世纪 80 年代以来,世界人均存储信息的技术能力大约每 40 个月翻一倍;截至 2012 年,每天生成 2.5EB(1EB=1024PB=1048576TB,1TB=1024GB)的数据。到 2025 年,每天预计将达到 491ZB(1ZB=1024EB,1YB=1024ZB)的数据。对于大型企业来说,确定谁应该拥有影响整个组织的大数据是一个迫在眉睫的问题。

大数据(图 6.17)是一个数据集,它数量庞大而复杂,一般传统的数据处理应用软件是不足以应付这些数据的。大数据面临的挑战包括数据获取、数据存储、数据分析、搜索、共享、传输、可视化、查询、更新、信息隐私和数据来源管理。与大数据相关的术语有很多,如 volume(大量)、velocity(高速)、variety(多样)、value(低价值密度)、veracity(真实性)。

近年来,大数据倾向于指使用预测分析方法、用户行为分析方法或其他高级数据分析方法,从数据中提取价值,并且很少用于特定大小的数据集。现在可用的数据量确实很大，但不是这个新数据生态系统最相关的特征。分析数据集可以发现新的相关性，以发现商业趋势、预防疾病、打击犯罪等。企业高管，医学、金融科技、城市信息学和商业信息学等领域的从业人员都遇到了大数据集的困难。科学家在科学研究信息化的工作中也遇到了挑战，包括气象学、基因组学、连接组学、复杂的物理模拟、生物学和环境研究等。

云计算为大数据提供了相对便宜的存储空间和计算资源。从本质上讲，云计算是计算机程序的一种外包。使用云计算，用户可以从任何地方访问软件和应用程序；计算机程序由外部方托管并驻留在"云"中。这意味着用户不必担心存储和电源等问题，他们只需享受最终结果。

图 6.17 大数据

本 章 小 结

通过本章学习，要求学生了解高级数控技术和智能制造技术的发展，包括开放式 CNC 系统、柔性制造系统、虚拟轴机床、虚拟计算机数控系统和 3D 打印技术，以及人工智能、智能制造技术、虚拟现实和大数据等；掌握数控技术常用词汇，如 open CNC system、modularity、extensibility、reusability、compatibility、ethernet、SPC、FMC、FMS、NURBS interpolation、virtual axis machine tool、virtual CNC、3D printer、artificial intelligence、machine learning、intelligent manufacturing、virtual reality、big data、cloud computing 等。

本章的知识和能力图谱如图 6.18 所示。

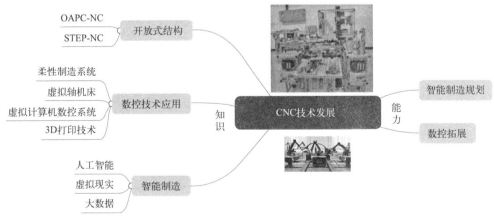

图 6.18 CNC 技术发展的知识和能力图谱

参 考 文 献

胡占齐，董长双，常兴，2004. 数控技术 [M]. 武汉：武汉理工大学出版社.

康兰，2016. CAD/CAM 原理与应用：英文版 [M]. 北京：机械工业出版社.

吴瑞明，2017. 数控技术：双语教学版 [M]. 北京：北京大学出版社.

ASTRI，2024. New industrialisation and intelligent manufacturing. [EB/OL]．[2024-07-19], https：//www. astri. org/technology/new-industrialisation-and-intelligent-manufacturing/.

Britannica，2024. Virtual reality．[EB/OL]．(07-18) [2024-07-22]. https：//www. britannica. com/technology/virtual-reality.

CIO Wiki，2022. Big data. [EB/OL]. (12-01) [2024-07-19]. https：//cio-wiki. org/wiki/Big＿Data.

ETHW，2023. Artificial intelligence [EB/OL]. (07-18) [2024-07-19]. https：//ethw. org/Artificial＿Intelligence.

OVERBY A，2010. CNC machining handbook：building, programming, and implementation [M]. New York：McGraw-Hill，Inc.

RAHMAN M，2004. Modeling and Measurement of Multi-axis Machine Tools to Improve Positioning Accuracy in a Software Way [D]. Oulu：University of oulu.

SUH S，KANG S，CHUNG D，et al，2008. Theory and design of CNC system [M]. London：Springer.

VALENTINO J V，GOLDENBERG J，2012. Introduction to computer numerical control [M]. 5th ed. Upper Saddle River：Prentice Hall.